BUTCHERING

BUTCHERING

BEEF

The Comprehensive Photographic Guide to
Humane Slaughtering and Butchering

Adam Danforth

PHOTOGRAPHS BY KELLER + KELLER

The mission of Storey Publishing is to serve our customers by publishing practical information that encourages personal independence in harmony with the environment.

EDITED BY Carleen Madigan
ART DIRECTION AND BOOK DESIGN BY Carolyn Eckert
TEXT PRODUCTION BY Theresa Wiscovitch

PHOTOGRAPHY BY © Keller + Keller Photography, except for pages v, 78, and 95 by Carolyn Eckert, and pages 70–73, 225, 253, 254, 255 (all except bottom), and 280 (left) by Mars Vilaubi.
ILLUSTRATIONS BY © Karin Spijker, Draw & Digit

INDEXED BY Christine R. Lindemer, Boston Road Communications

IN ADDITION TO THOSE MENTIONED IN THE AUTHOR'S ACKNOWLEDGMENTS, the publisher would like to thank the following people for their help: the Martin family at Elmartin Farm in Cheshire, Massachusetts; Jake Levin from the Berkshire Food Guild; everyone at The Meat Market in Great Barrington, Massachusetts; Brett McLeod; Shiloh Partin; Billy Barlow; Greg Stratton and Michael Montgomery from Stratton's Custom Meats & Smokehouse in Hoosick Falls, New York.

© 2014 by Adam A. Danforth

All rights reserved. No part of this book may be reproduced without written permission from the publisher, except by a reviewer who may quote brief passages or reproduce illustrations in a review with appropriate credits; nor may any part of this book be reproduced, stored in a retrieval system, or transmitted in any form or by any means – electronic, mechanical, photocopying, recording, or other – without written permission from the publisher.

The information in this book is true and complete to the best of our knowledge. All recommendations are made without guarantee on the part of the author or Storey Publishing. The author and publisher disclaim any liability in connection with the use of this information.

The publisher is not responsible for websites (or their content) that are not owned by the publisher.

Storey books are available at special discounts when purchased in bulk for premiums and sales promotions as well as for fund-raising or educational use. Special editions or book excerpts can also be created to specification. For details, please send an email to special.markets@hbgusa.com.

Storey Publishing
210 MASS MoCA Way
North Adams, MA 01247
www.storey.com

Storey Publishing, LLC is an imprint of Workman Publishing Co., Inc., a subsidiary of Hachette Book Group, Inc., 1290 Avenue of the Americas, New York, NY 10104

ISBNs: 978-1-61212-183-3 (paper); 978-1-61212-189-5 (hardcover); 978-1-60342-932-0 (ebook)

Printed in China by R.R. Donnelley
10 9 8 7 6 5 4

Be sure to read all instructions thoroughly before attempting any of the activities in this book. No book can replace the guidance of an expert, nor can it anticipate every situation that will arise. Always be vigilant when working with animals and use extreme caution when employing potentially lethal implements.

LIBRARY OF CONGRESS CATALOGING-IN-PUBLICATION DATA ON FILE

DEDICATION

For Mom, who fostered my love of food.

CONTENTS

Foreword viii
Introduction 1

CHAPTER 1
From Muscle to Meat 5

CHAPTER 2
Food Safety 21

CHAPTER 3
Tools & Equipment 34

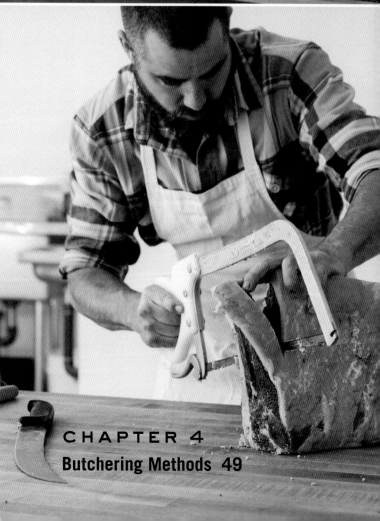

CHAPTER 4
Butchering Methods 49

CHAPTER 5
Pre-Slaughter Conditions & General Slaughter Techniques 75

CHAPTER 6
Slaughtering Cattle 93

CHAPTER 7
Beef Butchering 135

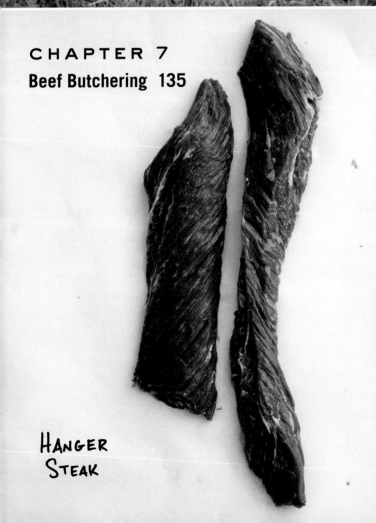

CHAPTER 8
Packaging & Freezing 311

Glossary 327

Resources 329

Index 330

FOREWORD BY TEMPLE GRANDIN

BECAUSE OF MY WORK DESIGNING SYSTEMS for humane slaughter, people often ask me about the ethics of eating meat. Early in my career, I thought about this constantly. One day I was standing on a catwalk overlooking pens of cattle that were going to die in the equipment I had designed. A flash of insight went through my mind: All the cattle that are going to be processed in this meat plant would have never existed if people had not bred and raised them. They would have never lived at all.

If we are going to raise animals for meat, however, we owe them a decent life and a respectful death. The Farm Animal Welfare Council in the United Kingdom has stated that an animal raised for food should have a life worth living. Cattle should have access to open land, clean water, and good food. When we harvest an animal for food, its death should be quick, painless, and stress free.

One advantage to butchering on the farm is that all the stress associated with transport and handling in a strange place can be avoided. An animal that dies when it is completely relaxed will provide the best meat; excitement, fear, and agitation that occur at the time of slaughter can toughen beef. Long-term stresses such as long transit time or cattle fighting before slaughter may cause dark cutting beef, a severe quality defect which raises pH and causes the meat to have a shorter shelf life.

There is a growing interest in small-scale processing of beef animals — partially because people are interested in maintaining control over the quality of their meat. Unfortunately, there's relatively little information that is appropriate for people operating at this level, especially those who are slaughtering for the first time. *Butchering Beef* by Adam Danforth provides easy-to-follow step-by-step instructions for people raising their own livestock to humanely slaughter a beef animal and butcher it with good food safety practices.

I remember vividly the day I killed my first steer. I had difficulty admitting to myself that I had actually done it. Even for experienced processors, though, killing should never become too easy. Taking the life of an animal should always be approached with respect.

Many cultures have slaughter rituals because they recognize that killing an animal is not the same as, for example, harvesting grain. Every animal we raise for food should have a life worth living, and every time we kill an animal it is our responsibility to provide a humane and painless death.

— Temple Grandin,
author of *Humane Livestock Handling*
and *Animals in Translation*

HUMANE SLAUGHTERING REMINDERS

- For both animal welfare and meat quality, it is essential that your beef animal be rendered unconscious instantly with a single shot from either a rifle or a captive bolt gun. Some people use feed to keep the animal still; while the animal is eating its favorite food, it is shot.

- Be sure to use a caliber of firearm or captive bolt that is appropriate for the size animal you will be slaughtering. A common mistake is to use a firearm or captive bolt that is too small. A .22 long rifle is the minimum required for calves, steers, and heifers. A rifle, due to spiral rifling of the barrel, will provide greater velocity (hitting power) compared to a pistol. Shotguns armed with slugs are preferred by many people. Bulls or bison require heavier firearms.

- If a firearm is used, the muzzle of the gun must be held a minimum of two or three inches away from the forehead.

- More powerful captive bolt guns are expensive but they are also more effective. If a captive bolt is used, I strongly recommend buying a high-quality gun from Bunzl-Koch Suppliers. A captive bolt must be held perpendicular to the animal's forehead.

- Captive bolt guns must be cared for the same way as the finest hunting rifle. After each use, they must be completely cleaned.

- Cartridges for both captive bolts and firearms should be stored in a dry place. Damp cartridges may cause a captive bolt to lose hitting power.

- For more helpful information, consult the guidelines of the American Meat Institute, American Veterinary Medical Association, the American Association of Bovine Practitioners, or grandin.com. – TG

INTRODUCTION

THERE IS NO BETTER LOCATION to harvest an animal than the land on which it lives. You're present with the animal, in a calm, familiar environment, during its final moments. Natural surroundings reinforce the normalcy of one animal's sacrifice for another's existence. Earth cushions the hard fall of a stunned animal, preventing bruising. Blood fertilizes the field.

Honorable harvesting prioritizes the well-being of the animal; the process resembles nothing of the horror stories — and horrific realities — coming from inside improperly operated abattoirs or industrial-size slaughterhouses. When I'm harvesting an animal, I feel right knowing that I have done everything I can to ensure a natural (albeit domesticated) existence and painless departure for any animal I process. This book will provide you with the same assurances.

Slaughtering an animal is not for everyone, and trepidation when beginning an education in this kind of work is to be expected. In fact, I would encourage it. A bit of uncertainty will make you slow down and will encourage you to deeply consider the importance of what is about to happen. Speed is for the machines. Act with intent and go slowly. Remember: preventing mistakes by practicing often and working with precision will increase efficiency more than working quickly will. At home there is no need to rush, either in raising the animal or in harvesting it. Slow growth yields a more flavorful meat, and cautious processing yields a nourishing product.

For people raising animals for their own consumption and for those looking to purchase and process animals that have been raised locally, this book is the key to your food freedom. Everything you need to know in order to successfully — and respectfully — slaughter the most common species found on a farm is contained in these pages.

You'll learn exactly how to prepare animals for slaughter, how to set up a slaughtering and butchering area, how to select the tools and equipment you'll need to ensure a successful slaughter, and, most importantly, how to stun and bleed animals with the certainty that they are experiencing the least pain and discomfort in these final moments of their lives.

Ensuring the animal's well-being during the slaughtering process isn't just about the ethics of humane animal handling; it's also about producing quality meat. Chapter 1, which focuses on meat science, explains exactly why providing better care, and slaughtering with respect, produces the best quality meat. When aberrations occur — an unsuccessful stun, punctured viscera, or damaged meat — there is comfort in knowing that you did what you could to avoid them. The slaughtering methods in this book focus on ensuring the animal is insensible to pain, not only for the animal's well-being but also for your own.

Knowing the ins and outs of slaughtering is just the first stage, though; the bulk of this book focuses on butchering the resultant carcasses. It behooves us to maximize the usage of each and every carcass we process, not only for our own return on investment but out of respect for these formerly living creatures. The many different butchering methods covered in this book ensure that you will find a system that works for what you want to eat and how you like to cook. Your favorite cuts are definitely in here, as well as many others that you've probably never encountered that might become new favorites. And, of course, there is detailed information on the ideal butchering setup for beef, along with the best options for equipment.

The butchering methods I demonstrate, although beef focused, can be applied to most any four-legged land animal. All animals share similar anatomical structures that have allowed us to saunter, trot, sprint, and exist outside of the oceans. Don't let the skeletal maps and scientific names of muscles send your mind spiraling; they're not critical for learning the skill of butchering. You will become a great butcher by committing shapes to memory, by understanding the intersection of muscles and the planes of connective tissue, by knowing the conformation of joints, and simply by making the same cuts over and over and over again.

An entire chapter is devoted to food safety and what we can do, as processors of a perishable and edible product, to ensure that the meat we produce is nourishing. The setup you devise at home may never be USDA certified, but the same rigorous dedication to sanitation and safe handling will benefit everyone who consumes the meat you produce.

It's deeply satisfying to know that, at my hand, an animal suffered as little as possible on its way to feeding myself and those around me. This kind of involvement also enables me

to produce meat that conforms to my standards: quality and cleanliness are paramount. When mistakes happen, during slaughter or butchering, I recognize them and make every effort to learn from such incidents. I don't rush to the end result, be it a carcass in the cooler or a roast coming out of the oven. I revel in the intricacies of a process that extends back in time for generations. Slaughtering and butchering my own meat connects me to cultures, regions, and generations I've never known. We all share the need to eat, and having a hand in how another living being is transformed and contributes to my own existence is profound and cathartic. I dearly hope that you draw the same inspiration from your experiences and that the information in this book helps guide you along the way.

Adam Danforth

FROM MUSCLE TO MEAT

PRIOR TO LANDING ON YOUR PLATE, the meat that you choose to eat began life as muscle, a highly organized and complex living system. Muscles, which are made up of tissues and fibers, perform many of the voluntary and involuntary actions in a body. Each muscle has a unique structure that depends not only on function and position but also on species and environment. As muscles are transformed into meat they undergo many physical and chemical changes. These changes are initiated by death but are influenced by many factors, including, for example, how the animal was handled before it was slaughtered and how quickly the carcass was cooled after slaughter. These factors and the chemical changes they cause can have an enormous effect on the palatability of the final cuts of meat.

As it turns out, the better you treat an animal while it is alive, the better the meat from that animal is. To create delicious meat, you should understand not only the physiology of muscles but also the types of favorable treatment that enable the production of a high-quality product.

Muscles

MUSCLES ARE ORGANIZED STRUCTURES that enable movement. The heart, a muscle, pumps blood through the body; muscles move food through the stages of digestion; muscles in the legs allow an animal to stand and walk. Each of these functions, enabled by the contraction of muscle, showcases one of the three different types of muscles: cardiac, smooth, and skeletal. But first, we must explore the basic muscle structure.

Muscle Structure

Muscles are made up of cells called *fibers*; these are slender cylindrical structures that enable contraction. Muscle fibers are organized in bundles that are stacked together in one direction and bound by sheaths of connective tissue. Envision holding a bundle of dry spaghetti; the spaghetti is the muscle fiber, and your hand is the connective tissue. This pattern of spaghetti-style bundling continues for many levels, as bundles upon bundles are grouped together, level after level, with the final bundle completing the full muscle. The connective tissue holding the full muscle together is the *silverskin* (technically called the epimysium). Along with more connective tissue, the space between the bundles is filled with blood vessels and fat deposits. The visual grain patterns we recognize in meat are actually midlevel bundles called *fascicles*. These are most notable in cuts in which the grain is prominent, such as the flank steak.

At the most basic level of the muscular structure are *sarcomeres*, long threads of linked proteins organized into bundles. These threads initiate muscle contraction from inside the muscle fibers. The main two proteins in sarcomeres, *myosin* and *actin*, make contraction and relaxation possible. They're linked in an overlapping pattern that allows them to slide past each other. When the muscle contracts they overlap more, shortening and getting closer together, and when the muscle relaxes they overlap less, lengthening the long threads they make. These protein actions change the shape of muscles; this is evident, for example, when you move your leg or flex your bicep. In short, when a muscle contracts, the action originates in the myosin and actin proteins. The action of these two proteins, shortening or lengthening, causes a chain reaction that repeats upward through every bundle: the fibers shorten, the fascicles shorten, and finally the entire muscle shortens for contraction, and vice versa for relaxation.

Muscle Fibers

Fascicles are the smallest muscle fiber bundle that we can easily identify with the naked eye and with the palate. The size of a bundle and its interior fibers plays a large part in how we experience meat. The larger the fascicle, the easier it is to see and the tougher it is to cut through. This gives us the advantage of being able to identify tenderness visually. Fine-grained muscles are more tender than coarse-grained muscles; thus a tenderloin is easier to chew than a skirt steak. The reason for this is that our teeth do a poor job of cutting through bundles of fibers; they are much more effective at separating them from one another. (Imagine trying to chop your way through a truckload of logs instead of just pushing the logs to one side or another.)

In addition, muscle fibers typically toughen during cooking, drying up as the heat ruptures water-holding structures and causes evaporation. The result is a denser, more resistant structure. This is the reason we cut meat across the grain rather than with it. Cutting with the grain would leave stacks of dense, lengthy fiber bundles that we would struggle to split with our teeth. Instead, we let a knife do the work of shortening the fibers so our teeth can do the job of separating them, an effort that in some cases takes 10 times less energy than splitting.

MUSCLE FUNCTION AND AGE DETERMINE SIZE

The size of muscle fibers is partly the result of muscle function. The more power a muscle needs, the shorter and fatter the sarcomeres within the fibers. More power requires more contractile proteins (actin and myosin). This in turn requires stuffing more proteins into the same connective tissue casing. (Imagine, for example, filling a balloon to capacity with water.) When more proteins are created to produce more power, this causes the sarcomeres to fatten. As sarcomeres fatten, so do the fibers, the bundles of fibers, and the bundles of bundles throughout the entire muscle. Muscles requiring short, powerful bursts of energy — such as those responsible for an animal's fight-or-flight response in reaction to sudden danger — have the thickest fibers. One example of this is the breast muscle of birds that fly only when threatened, such as chickens. (You may be saying to yourself, "But the breast meat of a chicken is so tender." Muscle fiber size is not the only factor in determining tenderness; see page 12.)

Age also contributes to muscle strength and therefore fiber size. In general, the older an animal is the larger it gets, and the longer it has lived the more activity the muscles have experienced. An animal does not grow new muscle fibers; rather, the fibers increase in size as they develop more contractile proteins. The muscles require more strength to support the growing size of the animal; as the animal ages, increased activity promotes muscle expansion. Larger muscles need more strength, provided by an increase in contractile proteins (actin and myosin). The more proteins inside a muscle fiber, the denser and wider the fiber, and the tougher it is to chew. This is one reason why older animals have tougher meat.

Connective Tissue

Connective tissue is made primarily of *collagen*, a substance that accounts for about one-third of the protein in the entire animal. Collagen is concentrated the most in ligaments, tendons, bones, and skin. The other notable component of connective tissue is *elastin*, which is named for its elastic properties and provides some of the stretch that connective tissue needs in order to change shape and move with the muscles and other body parts.

CONNECTIVE TISSUE AND STRUCTURE

The structure of all tissues within the body, muscles included, is enabled by connective tissue. Muscle fiber bundles, and the bundles of bundles, are all wrapped by thin layers of collagen-rich connective tissue. Within these bundles, numerous strands of connective tissue fill the spaces between fibers. These strands weave themselves together, as in a tapestry, to form a complex structure. The strands are connected through a process called *chemical cross-linking*. The interior and exterior networks of a muscle's connective tissue all converge at either end to form tendons. When muscles contract, fibers tug on their respective connective tissue sheaths,

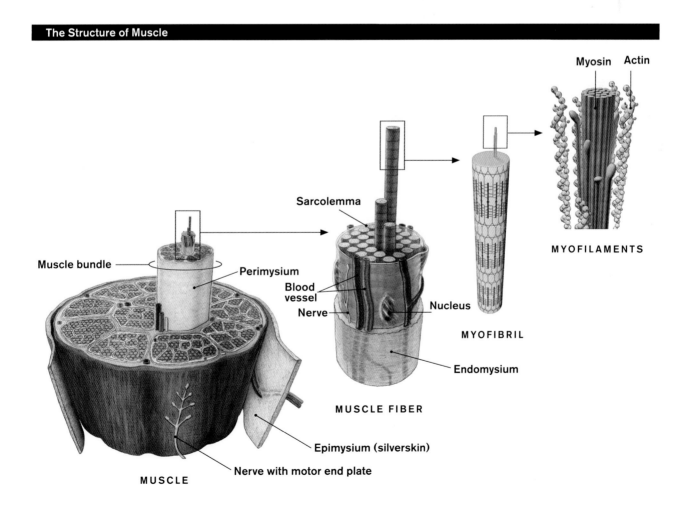

The Structure of Muscle

causing bundles of fibers to contract. Through a chain reaction across the bundles of bundles, the muscle pulls the tendons and causes skeletal movement.

COLLAGEN AND MUSCLE TENDERNESS

More than any other factor, the main property that governs muscle tenderness is the volume and strength of cross-links between collagen fibers. Just as with textiles, the more threads and connections you have, the stronger the fabric and the tougher it is to cut through. Many factors contribute to cross-link development, including not only the function of the muscle but also the animal's age, nutrition, and breed. The hardest-working muscles, and those that get the most exercise, require a dense network of collagen to provide adequate structure and functionality. Density is achieved through the development of intensely cross-linked collagen fibers. As a rule, the closer to the ground a muscle is, the harder it works to provide support to the body. This is illustrated in the copious amounts of collagen found in meat from the lower limbs of all animals, including beef shanks, ham hocks, and chicken drumsticks.

BREAKING DOWN COLLAGEN BY COOKING

Fortunately, collagen and its cross-links can be broken down into gelatin through the application of heat and water, a process called *hydrolysis*. *Gelatin* is the sticky, unctuous substance that helps thicken liquids for sauces or desserts and provides the adhesive for traditional glues. In contrast to muscle fibers, which get drier and denser when cooked, collagen softens during a proper stewing, helping to turn otherwise tough cuts of meat like a beef shank into a succulent result. As a general rule, the tougher the collagen and the more cross-links it has developed, the longer it will take to break it down into gelatin. Thus, meat from an older animal will need more moisture and time to hydrolyze than meat from a younger animal of the same species.

Hydrolysis of collagen begins as the temperature rises above 122°F. The higher the temperature, the faster it happens. However, while higher temperatures increase the rate of hydrolysis, there is a trade-off. Once the temperature rises above 140°F the collagen also begins to shrink. The shrinkage begins to squeeze on the muscle fibers, causing them to expel liquid. The process is similar to twisting a wet towel: the more you twist, the more water flows out. The higher the temperature, and the quicker it rises, the faster and tighter collagen strands twist and squeeze out the moisture contained in muscle fibers. Hence, hydrolysis that occurs too quickly results in dense, dry meat.

Take a beef shank, for example. A beef shank is heavily worked and therefore chock-full of extensively cross-linked connective tissue. Cooking this cut for a few hours at a high temperature, in moist or dry heat, will produce meat that is dense, dry, and a struggle to chew: the collagen has shrunk, squeezed out the liquid, and not been given adequate time to hydrolyze. Slow-cook it at a low temperature for many hours, and the muscle fibers will fall apart into tender threads of meat: the collagen has been fully hydrolyzed, and the structure holding the fibers together turned into gelatin. With a longer cooking process, the transformation of collagen into gelatin provides a better mouthfeel. Slow-cooked meat has still been squeezed by the collagen, though, so it will still benefit from the application of moisture, such as the reduced liquid in which the shank was cooking (which now contains generous amounts of gelatin, adding a pleasing mouthfeel).

Allowing time for hydrolysis is pertinent only when dealing with tough cuts of meat in which there is a substantial amount of connective tissue. Tender cuts have weak collagen and in small amounts. The generally preferred method for tender meat is quick cooking because there is not enough collagen present to make chewing difficult. Further, keeping the internal cooking temperature of tender cuts of beef to 140°F or lower avoids collagen shrinkage and the resultant moisture loss. This is why a tenderloin cooked to medium-well at 150°F or higher will be a denser, drier version of the same steak cooked to a 133°F medium-rare.

Diagram of collagen fibers and cross-links

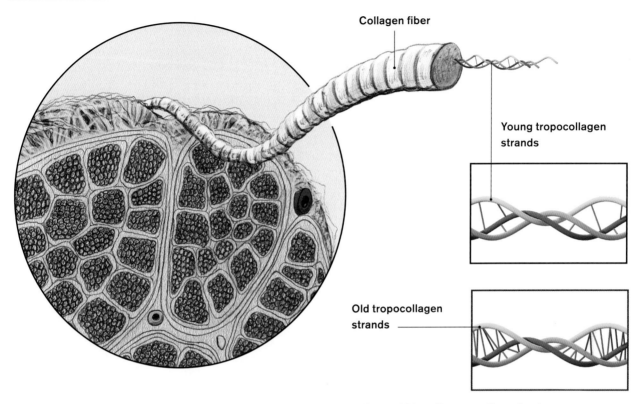

Tenderness is largely the result of cross-links between tropocollagen fibers within collagen; as the animal ages, more cross-links are formed, making the meat progressively less tender.

AGE AND ITS EFFECT ON COLLAGEN

As an animal ages, the volume of collagen decreases but the strength increases. Aging of the muscle causes the development of more chemical cross-links between the collagen fibers that remain. Muscles are exercised, fibers increase in density and girth, and the collagen fibers respond accordingly by increasing tensile strength through the addition of cross-links.

To avoid harvesting meat with tough collagen, those who raise animals for the meat industry slaughter most of those animals before they reach adulthood. For example, consider the difference between a four-month-old veal and a three-year-old beefer. The muscles throughout the veal chuck (the shoulder), with their relatively weak collagen cross-linking, are notably tender, allowing for more versatile preparations; the beef chuck, on the other hand, contains numerous muscles that display the characteristic toughness of strong collagen fibers, requiring long, slow cooking to break them down.

Fats

Along with fibers and collagen, fat plays a distinctive role in our experience of consuming meat. Fats happen to be a unique form of connective tissue, primarily serving three purposes, in some cases simultaneously: to insulate the body, to protect the body and the internal organs, and to store energy. The latter function is responsible for many of the flavors that we associate with meat. To increase fat coverage, an animal does not add new fat cells but rather increases the volume of the cells already there.

FAT CELLS AND TASTE

Fat cells store energy in the form of fatty acids but also act as a repository of any substance that is fat-soluble. (Just as salt is water-soluble, any compound or substance that will dissolve in fat is fat-soluble.) So, while an animal gathers energy from its food, it also stores other fat-soluble compounds from the food within the fat cells. Which compounds are stored depends largely on species and diet, while the concentration of those compounds is mainly a result of age. The older an animal is, the more time it has spent storing fat-soluble compounds in its fat cells and the more flavors and flavor-enhancing components are released during cooking. This accounts for the typically stronger flavor and aroma of meat from older animals, as is the case with mutton.

An animal that is raised primarily on pasture, relying on a varied diet of foliage, both fermented and fresh, will process and store a diverse array of organic compounds and fatty acids. Upon cooking, these assorted odorous substances will strengthen the flavor of the meat. In contrast, an animal reared with a diet composed primarily of grain will have less diversity in its fat stores. It is for this reason that meat from grass-fed and pasture-raised animals has a stronger flavor than meat from grain-fed animals.

SATURATED AND UNSATURATED FAT

Within the world of animal fats there are two main categories: saturated and unsaturated. Fats are composed of carbon atoms linked together in chains. These carbon atoms like to bind with hydrogen, and in the case of a satu-

TYPES OF FAT DEPOSITS

Within the body of an animal, there are four types of fat deposits.

SUBCUTANEOUS FAT lies under the skin, as seen on this beef round.

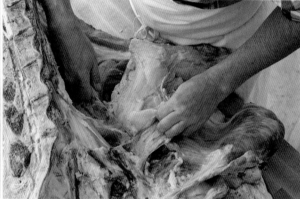

VISCERAL FAT like this kidney, or cod, fat lies inside the body cavity. It also surrounds kidneys and other organs (caul fat, for example).

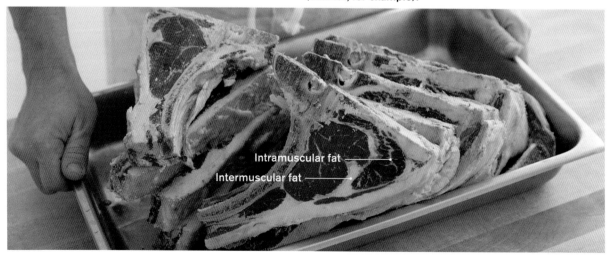

Intramuscular fat
Intermuscular fat

INTERMUSCULAR FAT is intermingled between muscles (as seen in these rib steaks).
INTRAMUSCULAR FAT, or marbling, is interspersed within the connective tissue and fibrous bundles of muscles.

rated fat, the carbon atoms bind with as many hydrogen atoms as possible. (They are literally *saturated* with hydrogen bonds.) Unsaturated fats are not saturated with hydrogen bonds. Instead, one or more carbon atoms are double-bonded to each other. Monounsaturated fats have a single double bond; polyunsaturated fats have more than one double bond. This double bond adds one or more kinks to the chain, changing its shape from clean and organized to a bit awry.

If you're good at organizing, packing a car, or stacking boxes, you know that things of a consistent shape fit tightly together. This is the case with saturated fats: the chains of evenly bonded molecules stack tightly together, forming stable fats that are solid at room temperature. Once there is a kink in the chain, those chains can't stack so closely, thus preventing them from forming tight, stable structures, often making them liquid at room temperature; the more kinks the more unstable the structure. Animal fats contain *mainly* saturated fats, making them solid at room temperature. But the amount of unsaturated fats certainly comes into play. Chicken and pork fat have higher levels of unsaturated fats than beef, sheep,

TYPES OF MUSCLES

The three types of muscles are defined according to both function and form.

CARDIAC MUSCLE is found only in the heart. It's striated and involuntary.

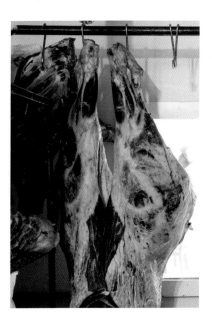

SKELETAL MUSCLE is attached to the skeleton and enables its movement and is the predominant type in the body, as well as in butchery. It's striated and voluntary.

SMOOTH MUSCLE comprises all the remaining involuntary muscles such as the walls of blood vessels, the gastrointestinal system, and the stomach lining. It's nonstriated and involuntary.

and goat fat, making them less solid at room temperature. (Vegetable fats, like olive and canola oil, are mostly unsaturated and that's why they're liquid.)

These kinks in the molecular chain affect not only the solidity of the fats but also the rate of rancidity. A double bond in the chain exposes carbon atoms to oxidizing elements, like oxygen and water. These elements react with the exposed carbon atom, often disrupting the chain, causing it to break apart. This is the basic action of rancidity or oxidation — the breakdown of molecular chains into smaller fragments. The more kinks in the chain, as with polyunsaturated fats, the more susceptible they are to rancidity. The more unsaturated fats in the animal, the more prone the carcass and its resultant cuts are to rancidity and oxidation. Because poultry and pork have higher levels of unsaturated fats, their carcasses can't be aged as long as beef, sheep, or goat carcasses can, and the cuts from these animals tend to spoil sooner, whether fresh or frozen. (Liquid fats have a tendency to go rancid even more quickly, especially nut oils that are high in polyunsaturated fats. This is why refrigeration of these oils is always recommended.)

FAT AND MEAT TENDERNESS

Fat is also a contributor to tenderness, but marginally so compared to fibers and collagen. Within muscles, the presence of intramuscular fat between bundles and fibers disrupts the mesh of collagen fibers and cross-links, making holes in the otherwise taut tapestry. This helps to weaken the solidity of a muscle's connective tissue, increasing tenderness when we chew. Unlike fibers, which dry out, fats also melt during cooking. Melted fats turn to liquid, adding a necessary moisture to the drying fibers. This also lubricates the fibrous bundles, aiding our teeth's efforts to separate them, resulting in a more tender bite. Additionally, as fats melt under heat they release the aromatic fat-soluble compounds stored within the cells, contributing to the olfactory experience and increasing the perception of flavor.

Types of Muscles

The function of a muscle can be either *voluntary* or *involuntary*: voluntary contractions are performed with intention (the movement of an arm or the focusing of eyes), whereas involuntary contractions happen without conscious control (the drawing in of air to breathe or the pulsing of a blood vessel). The form of a muscle can be *striated* or *smooth*. Striated muscles contain parallel fibers, organized side by side, providing the grain and appearance we associate with meat. Smooth muscles have fibers organized in complex sheets and are associated with blood vessels, organs, and other internal functions.

Types of Muscle Fibers

The main function of muscles is movement, and animal movement can be separated into two basic types: quick bursts or slow and steady. The different types of movement are achieved by two types of muscle fibers: a fast-twitch fiber, which provides the sudden contraction needed for bursts of energy; and a slow-twitch fiber, which provides the endurance essential for sustained activities like standing or chewing.

FAST-TWITCH FIBERS

Fast-twitch fibers are strong, as their contractions require power to achieve the appropriate speed of action. They are short, fat, and hard to chew, in accordance with the aforementioned effects of strength on fiber size. Moreover, fast-twitch fibers do not need the energy from fat stores to operate, making the muscles dominated by these fibers very lean. Yet fast-twitch muscles are called into action only periodically, which makes for a sparse and weak network of collagen. So while the fibers are harder to chew, with meager fat, the muscle overall is in most cases considered relatively tender. A well-known example is the chicken breast, a tender, lean muscle composed of fast-twitch fibers that illustrates how connective tissue is the most influencing factor in tenderness.

SLOW-TWITCH FIBERS

Slow-twitch fibers are long and thin, requiring less power but more endurance than fast-twitch fibers. Their stamina is provided through a combination of extensive connective tissue and large intramuscular fat stores. The connective tissue provides the rigidity and support required to sustain lengthy periods of contraction and activity, while the fat stores provide the energy necessary to operate for extended periods. Muscles with slow-twitch fibers are normally considered tough. Despite the tender narrow fibers and the generous fat stores, the abundant connective tissue is the determining factor creating chewiness. Hardworking muscles in the shoulders of quadrupeds or in the legs of poultry are typical examples of how the concentration of connective tissue determines tenderness. Under the proper cooking conditions these muscles provide some of the most unctuous results. The exception to this is a muscle like the tenderloin. It is a rare combination of typical slow-twitch muscle fibers, but since it is rarely used in the body of quadrupeds, it has very little connective tissue, making it the most tender of all cuts.

Each muscle in the body is a mixture of fast- and slow-twitch muscles. Fast-twitch muscles are distinctly paler than slow-twitch muscles, resulting in varied colorations within the muscles of an animal. An easy example is the typically white chicken breast and the dark-colored thighs. The breast, comprised mostly of fast-twitch fibers, explodes with energy when the chicken takes a sudden but short flight. The legs, which are slow-twitch dominated, require the stamina for daylong posture and movement.

As a general rule, muscles nearer to the skin are expected to respond less frequently than those farther away and are therefore lighter in tone and leaner in composition. Muscles residing closer to the skeleton are the ones handling the bulk of the posturing and movement; these muscles will be darker in tone and contain higher concentrations of slow-twitch fibers, connective tissue, and fat. So, while that chicken breast may be lightly colored, lean, and tender, the absence of connective tissue and fat means that it can easily be turned into a dense, dry cut, a lackluster comparison to a properly cooked chicken leg in which the collagen has been hydrolyzed into gelatin and the fats have flooded the meat with fragrant and tasty compounds.

The Color of Muscles

The coloration of muscles does not, as some people think, come from blood. In fact, there is no blood within the muscle fibers themselves. Muscles do need blood to operate, though, and vessels running among the connective tissue and fat that sits between the fibrous bundles carry blood into the muscles. Muscle fibers need oxygen, and to get it from the bloodstream to the fibers requires a local delivery system of sorts. In comes *myoglobin*, a protein capable of making the trip between the bloodstream and the muscle fiber while carrying a load of oxygen. While an animal is alive, oxygen exchanges are constantly happening according to the demands of a muscle. The more a muscle works, the more oxygen it needs and the more myoglobin is there to make it happen.

Beef chucks and other slow-twitch fibers are deep in color and require connective tissue and fat deposits for endurance.

THE THREE STATES OF MYOGLOBIN

Myoglobin happens to be the main party responsible for muscle coloration and is able to change color according to its state. Myoglobin has three states: deoxygenated (not carrying oxygen), oxygenated (carrying oxygen), and oxidized (exposed to external elements).

DEOXYGENATED. This first state occurs when myoglobin is not holding on to any oxygen. After an animal is dead there is no more circulation of oxygen-rich blood and no need for muscle fibers to operate: no more oxygen for the myoglobin to carry around. So the default state of myoglobin in meat is the deoxygenated state, the result of the myoglobin's last delivery of oxygen to an eager muscle fiber. Meat in this first state is purplish-red. This may sound familiar to anyone who has sliced open a raw steak or separated a bunch of ground beef to find the interior tinged with a purple hue; red meat cuts that have been vacuum sealed also tend to show deoxygenated coloration. What you see is myoglobin without access to oxygen — that is, until you give it access to the air around it.

OXYGENATED. The second state of myoglobin occurs when it picks up some oxygen. When myoglobin is oxygenated it is called oxymyoglobin, which is bright red. Oxygenation can happen from anywhere oxygen resides: living muscles gather oxygen from the bloodstream; meat picks up oxygen during exposure to air. Myoglobin binds with oxygen whenever it has the opportunity, so when you cut open that raw steak, break apart a pile of ground lamb, or open a vacuum-sealed packaged of red meat, the hue shifts from purplish-red to cherry red, a process known as "blooming."

OXIDIZED. Myoglobin may like to carry around oxygen, but it is not all that great at holding on to it. The bond between myoglobin and oxygen is highly unstable and susceptible to being broken through exposure to things like bacteria, enzymes, light, or even more oxygen. When oxygen breaks away, it often steals an electron, leaving myoglobin in an oxidized state called *metmyoglobin*. Metmyoglobin is brownish-red and responsible for the unattractive hue of meat left in compromising conditions for too long — on the counter, in the fridge, under heat lamps — but it does not always indicate spoilage or rancidity.

No muscle is exclusively one type of fiber, so the shades of muscle color are a result of the mixture of slow- and fast-twitch fibers. This is easier to identify within sedentary animals, which have higher concentrations of fast-twitch fibers and intermittently used muscles, making the darker hues more evident. Pigs are

DEOXYGENATED. The absence of oxygen within a vacuum-sealed package results in the purplish hue of deoxygenated myoglobin.

OXYGENATED. Red meat "blooms" to its characteristic red coloration as myoglobin bonds to oxygen in the air.

OXIDIZED. Extended exposure to the elements results in metmyoglobin, causing red meat to shift to an unattractive brownish-red hue.

Cured meats retain their attractive red hue because of a strong bond between myoglobin and nitric oxide.

not grazing animals and move much less than cattle do; hence, pork is lighter in color than beef. Within grazing animals and other animals whose bodies are in constant motion, a third type of muscle fiber develops: the intermediate fiber. The intermediate fiber is a fast-twitch fiber with aerobic capability; it therefore needs oxygen and contains myoglobin. A combination of intermediate and slow-twitch fibers mainly composes the muscles of red meat animals like cows, lambs, goats, and deer. It is this overall coloration of red meat that makes differentiation between slow-twitch and fast-twitch fibers difficult to discern in many species.

The size of the animal and demands for movement also account for differences in the volume of myoglobin among species. Beef, which comes from large animals that cover a great deal of ground, has more myoglobin than either pork or lamb, making it exceptionally red.

Finally, as an animal ages the volume and the concentration of myoglobin increase, resulting in richer coloration.

COLOR AND THE FRESHNESS OF MEAT

The final color of meat is a mixture of purple, red, and brown — the colors of myoglobin molecules in all three states. Shifts between states, caused by oxygen handling and electron theft, are constantly happening, but the dominant condition will determine the color. A vacuum-sealed beef steak will be purplish because there is not much oxygen inside the bag. Open it up, though, and the steak will bloom to red within a few minutes as the myoglobin is exposed to the oxygen in the air. If the meat is left out longer, the myoglobin loses hold of the oxygen and turns brown. We have come to associate color with freshness, as with the bright red of beef, but in many instances the presence of brownish hues is not an indication of rancidity or spoilage.

PRESERVING THE COLOR OF MEAT

There are two conditions in which we intentionally interfere with myoglobin to produce desirable colors under undesirable conditions: when we cure meat, and when fresh meat is packaged for sale.

CHEMICAL PRESERVATION. Cured products often use nitrites, a chemical added to meats to prevent bacterial growth and control the final color. During the curing process, nitrites convert to nitric oxide, a compound that can bind with myoglobin, effectively taking the place of oxygen. This turns myoglobin pink and is why bacon, hams, and other cured meats have a pinkish hue. The myoglobin's bond with nitric oxide is much more stable than its bond with oxygen, so the color has better staying power.

PRESERVATION WITH CARBON DIOXIDE. Bright red meat sells better than brown meat does, but keeping it red is a challenge. (Remember, myoglobin loses oxygen easily and turns brown.) To keep meat red, processors pump carbon dioxide into packages of meat for sale. Carbon dioxide reacts with myoglobin in the same manner that nitric oxide does: it replaces oxygen, forms a stable bond, and turns the myoglobin reddish-pink. The color will not be as pronounced as the cherry red of the true oxymyoglobin, but it will certainly look better than the metmyoglobin brown and is therefore able to stay appealing in a meat case for a longer period.

Turning Muscles into Meat

TRANSFORMING MUSCLES INTO MEAT is a complex process that involves more than just proper slaughter and butchering. Fortunately, the same conditions that are most humane for the animal also produce the highest-quality meat products. How the animal is reared, fed, handled during transport, and treated prior to slaughter all play significant roles in the resulting meat. So do many postmortem conditions like storage, humidity, temperature, and especially time.

The effect of death on a body is nothing short of epic. The sudden standstill of the circulatory system halts all oxygen delivery, leaving aerobic cells without sustenance and causing immense cellular damage and chemical transformation. Enzymes and bacteria freely roam the defenseless carcass, mercilessly breaking down structural components. Fatty and amino acids are left to disintegrate. It may seem chaotic, but under proper conditions this natural process transforms previously living tissue into palatable meat.

The First 24 Hours Postmortem

The events that occur in the first 24 hours after death can determine the quality of all meat coming from a carcass. Thankfully, the conditions that can cause those events are largely within our control. Living muscles are constantly using energy, typically in the form of glycogen. When muscles process glycogen, the by-product is lactic acid. The more a muscle works, the more lactic acid is produced. (Lactic acid is responsible for the muscle burn associated with intense exertion.) Oxygen from the circulatory system maintains the level of lactic acid, removing it when too much accumulates. After death, muscles continue to generate lactic acid, but with no blood circulation the acid builds up, changing the pH level from a nearly neutral state (around 7.0) to a slightly acidic state (near 5.7). This is beneficial, as the mild acidity retards microbial activity, including that of enzymes and bacteria. In addition, the drop in pH causes proteins to slightly unravel (a process called denaturing), releasing some fluid and further moistening the meat.

RIGOR MORTIS

Muscles contract and relax in response to chemical reactions. Under normal conditions, the drop in pH is a slow decline over the course of many hours. Muscle fibers continue to contract and relax after death, but as the pH drops the chemicals that allow the fibers to relax become sparse. Eventually the muscles contract and never relax again, causing the stiffening known as *rigor mortis*. There are three phases to rigor mortis:

- **THE DELAY PHASE** — muscles continue to contract and relax after death.

- **THE ONSET PHASE** — muscles begin to lose the ability to relax and start permanently contracting.

- **THE COMPLETION PHASE** — all muscles have tightened to a fully contracted state.

Rigor mortis effectively exhausts muscles of their ability to contract and relax; after rigor, muscle functioning is completely finished. But, to become palatable meat, more time is needed and more chemical reactions must take place.

The effects of rigor mortis are counteracted by enzymes and time. Once an animal dies, enzymes are let loose, and with nothing regulating their activity, they begin to attack at random. Their targets include the contractile proteins that keep muscle fibers in a contracted state (remember actin and myosin from page 6). Given enough time, these enzymes will have done enough dismantling to undo the effects of rigor mortis, producing tender meat. This is all part of the process of aging, discussed in detail below.

The Effects of Animal Stress on Meat Quality

Although normal changes in pH levels are advantageous, aberrations in pH levels can cause unsavory effects. Stress is the main culprit. Animals exposed to stressful conditions for extended periods prior to slaughter (e.g., uncomfortable temperatures, jarring noises, unstable surfaces for standing, confrontations in the pen) will become exhausted. After they die, their muscles will not produce enough lactic acid to induce the proper drop in pH levels, resulting in a high final pH (above 6.0, depending on the species). Meat with such a high pH is referred to as dark, dry, and firm (DDF) or "dark cutter." Without the acidity needed to curb microbial activity, spoilage quickens: proteins tighten, rather than denature, and strengthen their bond with water, causing the firm texture and dryness. Red meat is especially susceptible to DDF, with most cases occurring in beef.

Similarly, a sudden fall in the pH of muscles after death also results in poor-quality meat. When an animal is threatened, surprised, or excited its fight-or-flight response results in a flood of adrenaline entering its muscles. If this happens within a half hour prior to slaughtering — for example, because of fighting with other animals, being mishandled, or experiencing loud noises on the kill floor — the effects of the adrenaline will continue to persist after death.

An adrenaline rush spurs accelerated processing of glycogen and increased lactic acid production. A sudden shift of pH from neutral (7.0) to acidic (less than 5.8), when combined with warm muscles, causes the muscle fibers to unravel excessively. The effects are numerous, but the outcome is called pale, soft, and exudate (PSE) meat. In short, it's wet, mushy meat that dries out easily. As the muscle fibers unravel, they lose much of their ability to hold water, resulting in excessive moisture or drip loss and a compromised structure; at the same time, the myoglobin changes in structure and reflects light, causing a paler-than-normal color. PSE meat caused by pre-slaughter stress can come from any animal, but the most common occurrences are in pork because of the genetic disposition of pigs.

The Importance of Proper Storage after Slaughter

Production of high-quality meat requires careful planning and attention through each stage of life, slaughter, and storage. Even when all preceding conditions are carried out to the ideal, improper storage procedures have the potential to compromise the final quality of meat. Carcasses are chilled soon after slaughter to prevent the growth of dangerous microbes that can cause spoilage and illness; the quicker the chill, the less bacterial growth.

Chill the carcass too quickly, though, and the consequence is incredibly tough meat caused by a process called *cold shortening*. This occurs when the temperature of the meat drops below 59°F before the onset phase of rigor mortis. At temperatures below 59°F, muscles contract to abnormal extremes. This causes shortening of the muscle fibers (and therefore the muscles), in some cases to less than 50 percent of the original length. Therefore, when the final contraction of rigor mortis sets in, the result is a magnified tightening of the muscles, causing irredeemable toughness of the meat.

A similar condition, called *thaw shortening*, happens when meat from a freshly slaughtered carcass is frozen prior to the onset of rigor mortis. In this case the meat never goes through rigor mortis, therefore never exhausting the muscles of their ability to contract. When the meat thaws, the muscles come back to life (in a manner of speaking), contracting in the same manner as with cold shortening and ending with similar results.

> The same conditions that are most humane for the animal also produce the highest-quality meat products.

Aging

PALATABLE MEAT IS THE RESULT of properly aging the carcass to counteract the effects of rigor mortis after the animal has been responsibly slaughtered. There is a sliver of time after slaughter in which the muscles are still relaxed and, if cooked, may provide a tender bite. This period is impossibly short for small animals like chickens and rabbits; for larger ones, the result of cooking the meat at this stage would be a watered-down, mildly flavored version of what you would expect. Soon after this sliver of time, rigor mortis sets in; attempting to butcher, cook, freeze, or do anything else with the meat during the period of rigor mortis would result in total failure. At this point the opposing muscles (e.g., hamstring/quadriceps, bicep/triceps) in the carcass are contracting, ending in the unusual posture associated with stiffened roadkill. The contractile proteins actin and myosin have permanently bonded to each other. This is where aging comes into play: meat will tenderize and improve itself given the right conditions and proper time.

Enzymes at Work

Aging does many things, but its main effect is to undo the results of rigor mortis. After death, naturally occurring enzymes within the meat go rogue, since the body systems that kept them in line shut down. The main agents are two types of enzymes called *calpains* and *cathepsins*. They indiscriminately attack proteins and, in a process called *proteolysis*, break them down into fragments, disrupting the structures responsible for keeping muscle fibers in a contracted state.

Given time, the enzymes dismantle enough contractile structures in the muscle to undo the structure of contraction and the effects of rigor mortis, essentially relaxing the muscles and increasing tenderness. Moreover, the cathepsins take apart cross-links and fibers within the connective tissue, another benefit for tenderness. The result is collagen that hydrolyzes more easily during cooking, producing more gelatin and thereby increasing succulence. The weakened collagen structures also prevent excessive squeezing on muscle fibers during cooking, stemming excessive moisture loss. The by-product of all this enzymatic tenderization is a wide range of broken-up proteins and molecules: a great thing for our palate. Formerly bland fats, proteins, and other molecules are all transformed into fragmented and intensely flavored compounds. These account for many of the sweet, savory, and aromatic attributes associated with aged meat.

The longer a butcher waits to process the meat, the more structures collapse from enzymatic activity and the more tender the product. The meat is quite literally rotting, but under very controlled conditions. The bustle of enzymes increases with temperature, doubling every 18°F, but the development of harmful microbes also spikes. So, while we could speed up aging by raising the temperature, it behooves us to bide our time and keep the meat between 32°F and 38°F. At these temperatures, the enzyme mayhem continues, albeit sluggishly, and the meat is safe. If the temperature of the meat goes below 28°F, which runs the risk of freezing, proteolysis slows down to an inconsequential crawl.

How Long to Age

All animals benefit from controlled aging to counteract rigor mortis, though the length of the aging process is different for each species and is further influenced by ambient temperatures, air flow, humidity, and personal taste. General rules are that smaller animals age more quickly than larger ones and younger animals age more quickly than older ones. Beef should be aged 10 to 28 days. Middle-meat primals — rib, short loin, and sirloin — can be aged for several additional weeks, according to personal preference. The decision to age anything for more than 28 days, however, will be based on flavor and texture development rather than tenderness, as the enzyme action is minimal at that point.

Aging in the Open

All aged meat will increase in tenderness, but there are other beneficial repercussions, depending on airflow and the ambient humidity of where the meat is stored. In one method, called *dry-aging*, water evaporates from the meat, sometimes reducing the original weight by as much as 20 percent. With the water gone, the muscle fibers shrink, and so does the overall size of the meat. This also concentrates the tasty, water-soluble protein fragments, strengthening the flavor of the meat.

During the dry-aging process, meat is kept at the proper temperatures while humidity and airflow are controlled. Humidity is held at 70 to 80 percent, allowing the meat to dry out gradually. If the humidity is too low, the meat will lose moisture too quickly, resulting in dried, unpalatable meat; if the humidity is too high, moisture remains on the meat surface, promoting rancidity and microbial development. Air circulation is also critical to maintaining humidity equilibrium and promoting evaporation. To allow air access to all parts of the meat, meat processors usually hang carcasses from rails and place cuts on perforated shelves, while high-velocity fans work to keep the currents continuous.

A dry-aged carcass or primal cut will have a hardened, blackened exterior that is very likely to be dotted with patches of white mold. All mold patches must be removed and discarded, exposing the underlying nutty, aromatic meat. Between the loss of meat from trimming and the loss of weight through evaporation, the edible portion of a dry-aged primal may be 70 percent of its original weight. This makes dry-aging an expensive process: it requires equipment, ample space, and lots of time for hanging and trimming, and it ends with a considerable loss of salable weight. Yet the result, with its unique taste, will fetch high prices and yield flavorful results, making up for the product loss.

Aging in a Bag

These days, dry-aging is rarely done within the commercial meatpacking industry. Carcasses are typically hung for the minimum amount of time to allow rigor mortis to resolve, after which they are broken down into primal cuts. Primals are then vacuum-packed and shipped to customers within refrigerated containers.

While the meat is bagged and in transit, the enzymes do the work of dismantling proteins and tenderizing muscle. Upon arrival to a customer, bagged primals can continue to be stored and aged further or butchered into cuts. This approach is called *wet-aging*, and the results in tenderness are pretty much the same as in dry-aging. In wet-aging, the meat is aged within a hermetically sealed environment, staving off microbes and preventing any oxidation or moisture loss from the product. There is no need for controlled humidity or airflow, just temperature, so equipment and space costs are lower. Furthermore, there is minimal loss of product, thus maximizing salable weight. For these reasons, the adoption of wet-aging has been widespread within the commercial industries.

The downside is the resulting lack of flavor enhancement or development. The meat ages in a bag, spending days or weeks sitting in a collection of its own juices and blood. It picks up the flavors of these juices and blood: *serumy*, *metallic*, and *irony* are all adjectives used to describe the profile of wet-aged meat. Despite this, the benefits of minimal weight loss and convenient handling have made wet-aging the standard in modern meatpacking, an industry focused on speed and volume.

> The longer a butcher waits to process the meat, the more muscle structures collapse from enzymatic activity and the more tender the product.

Hanging carcasses in a temperature-controlled environment allows for extended aging.

FOOD SAFETY

THERE ARE MANY POSSIBLE REASONS that may have motivated you to begin slaughtering animals — animal welfare, cost reductions, custom cutting, to name a few — but one you may not have considered is the reduction of exposure to "industrial-strength" pathogens. Increased amounts of antibiotics in livestock, used for a myriad of reasons including promoting growth and preventing infections (rather than treating infections), have given rise to new strains of pathogens. Many of these new strains are showing resistance to the drugs we use to combat them. As the pathogens move from animals to humans, through our consumption of infected meat, their drug resistance poses a serious risk to you and all of us in the general public.

Pathogens and Contamination

THE NEED TO MEET the demand for feeding billions of people has motivated agriculture industries worldwide to rely on more and more automation and technology. The rise of automated systems has made the potential for spreading foodborne pathogens greater than ever. Ground beef, one of the products most susceptible to contamination, serves as a good example. A single pound of the stuff, sourced from your local supermarket, may come from hundreds (if not thousands) of different animals. Within these types of systems a single contaminated animal can cause widespread havoc. The consumption of such contaminated foods leads to infection, and the virulent strains responsible for these infections are becoming harder and harder to combat as more and more show signs of drug resistance.

Cleanliness during processing is paramount, not just for industrial meatpackers but also for home butchers. As butchers, we can help prevent transmission of foodborne pathogens, starting with mindfulness toward cleanliness. More than half of all foodborne illnesses in the United States result from contamination within the household, so it behooves us to take the appropriate steps to prevent contamination at home and in our work. Education and awareness are key to food safety, and as a home butcher you must make a commitment to reducing the potential for contamination.

The most fundamental concept to understand about food safety is that all food, without exception, is contaminated. No food is naturally sterile. No matter what we do, everything we eat is crawling with microbes. It just so happens that most of those microbes are not harmful to us, and the ones that are harmful may not be pervasive enough to overcome average immunity defenses. Our job, as butchers, is to do whatever we can to reduce contamination. The most powerful measure we can take is also the easiest: washing our hands. We'll cover more details later, but in essence, if you do nothing else to stem contamination (though you should take other steps as well), wash your hands; wash them often and wash them thoroughly.

The vast majority of foodborne illnesses stem from contamination caused by the people who have handled the food. For meat, contamination starts at the slaughterhouse. In most cases the muscles of an animal can be considered sterile prior to handling. With few exceptions, contamination is limited to the surface of the meat. The interior of a muscle is sterile until it's punctured or ground. The moment we cut into a muscle, we introduce surface contaminants; this may occur during cutting, tenderizing, probing for temperatures, or other kinds of meat handling. Simply put, everything we do escalates the level of contamination in the food. It is up to us to minimize this escalation.

Microbes, Good and Bad

Microbes are everywhere: on our desk, on our food, on our skin, even existing symbiotically inside our bodies. Most microbes are benign; some are even beneficial, like those residing in our digestive tract that aid in nutrient absorption. Microbes that pose a health risk to humans are considered pathogens (i.e., disease-causing agents); diseases contracted from food are caused by foodborne pathogens. Some pathogens can cause infection with only a single entity, as is the case with parasitic worms, while others require a colony of microbes to make an impact. We often consume small amounts of foodborne pathogens, mostly bacteria, but the immune system of a normal human is quite capable of handling these minor invasions. And although all people are vulnerable to contracting serious illness through foodborne pathogens, it's those with weak or compromised immune systems — infants, children, the elderly, and those with immune disorders like AIDS — who hold the highest risk of infection.

Foodborne Pathogens

Although there are five main types of foodborne pathogens (see page 24), the types can be split into two categories according to how they infect people: invasive and noninvasive.

INVASIVE INFECTIONS are caused by pathogens that penetrate the walls of the intestines, the stomach, or other areas of the digestive tract. The invasive microbe may also produce toxins or other means of infection. All five pathogens have a species capable of invasive infection.

NONINVASIVE INFECTIONS result from the body's reaction to toxins produced by the pathogen. These toxins may be excreted while the pathogen is inside our gut or may be produced in the food prior to our consuming it. As an example, take food poisoning, a seemingly catchall term for foodborne illnesses. In food poisoning, rather than waiting to produce toxins until it is inside our body, the food is adulterated with copious amounts of toxins before we eat it. It is for this reason that the onset of food poisoning is quicker than other conditions — the bacteria have already produced their toxins and therefore do not have to multiply and produce poisons while residing in the gut.

Causes of Contamination

There are many ways for food to become dangerously contaminated, but the most common involves either human feces or bodily fluids. Poor hygiene is the main culprit in the spread of foodborne pathogens. Food handlers (especially those in food service positions) who do not maintain strict hygiene habits have incredible potential for promoting widespread infections, often unintentionally and without awareness. An inadvertent cough or sneeze can spread millions of microbes. When produced by someone working near a vat of ground beef, a single sneeze can send microbes nationwide within distributed packages.

Animal feces are also a potent contaminant. Slaughterhouses are the most common site for contamination of meat via animal feces. The interior cavity and musculature of an animal is sterile, but the hide and feet abound with microbes. Because of this, removing the hide cautiously and hygienically is the single most important step to achieving a clean carcass. Contaminated meat can cross-contaminate any other meat it comes in contact with, which is why the high-volume operations of modern meatpackers carry an extreme risk of large-scale contamination: one bad apple will spoil the bunch (and anyone who consumes it).

Conditions for Contamination

In most cases pathogens need to multiply in order to be effective, and for them to multiply the conditions need to be right. After we ingest foodborne pathogens, there is a period in which they multiply inside our gastrointestinal system, building their population large enough to produce symptoms of infection. This period between ingestion and infection is called the *incubation period*, and the length of time can be dramatically different depending on the pathogen. For some bacteria, incubation can take as little as a couple of hours; for hepatitis, a common viral foodborne pathogen, the average incubation period is 28 days.

While many pathogens operate uniquely, they generally reproduce under similar conditions: warm temperatures, ample access to nutrients, and a slightly acidic pH. Both cooked and raw foods can provide the right nutrients and an acceptable pH condition, so the real variable for microbial growth on food is temperature. The range of temperatures that pathogens can actively persist in, often referred to as the "danger zone," is between 40°F and 140°F. (It is probably no coincidence that most pathogens multiply exceedingly fast at the same temperature as our body, 98.6°F.) Changes in the population of a pathogen can happen surprisingly quickly given the right conditions.

> The vast majority of foodborne illnesses stem from contamination caused by the people who have handled the food.

CONTAMINATION BEYOND MEAT

It is often a misconception that meat is more susceptible to pathogenic contamination than other food. The main issue with contamination is feces, and feces can be anywhere. Vegetables carry a high risk. In fact, some of the largest outbreaks of foodborne illness in the United States have come from vegetables. Farms may inadvertently spread infected manure as fertilizer on crops, leading to surface contamination on produce. An infected wild animal defecating in a field can also be a source of infection. In the home, proper washing of produce is paramount, especially for salads and other raw preparations.

Bacteria, for example, multiply exponentially (see Exponential Growth of Bacteria, page 25).

A general food service industry rule is that any animal protein exposed to temperatures within the danger zone for a cumulative amount of four hours should be considered unsafe for consumption. Unfortunately, it is rarely that simple. There are other influencing factors, but the higher the ambient temperature, the quicker the growth of microbes. Four hours may safely apply to temperatures on the lower end of the danger zone, but leaving a perishable product out in 100°F heat for four hours is a recipe for rancidity. As a specific example, leaving deli meat on the counter while you eat lunch during the winter may be fine. Leave it out a few times over the course of a hot summer week, and you'd probably want to throw it out instead of feeding it to your children.

Once meat is spoiled, nothing can bring it back to a state that is safe for consumption. Therefore, we take measures to prevent contamination early on, to thwart not only the microbes that cause spoilage but also the proliferation of pathogens. Holding food at a temperature below 40°F, the lower threshold of the danger zone, by refrigerating or freezing is one method. Cooking food to a temperature above the upper threshold of the danger zone is another way to prevent the growth of dangerous microbes. While you may aim for an internal temperature lower than 140°F for a ribeye steak, the exterior is undoubtedly heated well above that, making it impossible for surface contaminants to grow.

Types of Pathogens

FOODBORNE ILLNESSES ARE THE RESULT of an infection from one of five types of pathogens: viruses, bacteria, protists, parasitic worms, and prions. Incidents in the United States are common; estimates show that 9.4 million people become ill every year through foodborne contraction of one or more pathogens. The most common type is viruses, responsible for nearly 5.5 million cases of foodborne illness (59 percent). Bacteria account for about 3.6 million cases (39 percent), and protists and parasitic worm infections for around 200,000 (2 percent). Prions, while considered a foodborne pathogen, have yet to be specifically linked to any cases of human illness. While most cases are easily remedied, around 2,700 Americans die every year from foodborne illnesses. Though they account for less than half of the cases of illness, bacteria are by far the leading culprit in the deaths, claiming more than 65 percent of the lives lost. Protists — specifically *Toxoplasma gondii* (discussed further on page 28) — are responsible for another 25 percent or so of the deaths. Viruses, while being the largest infector, are the least deadly, causing 12 percent of foodborne deaths.

Bacteria

Bacteria, causing the most deadly forms of foodborne illnesses, catch all the headlines. Most people have heard of *Escherichia coli* (commonly referred to as *E. coli*) and salmonella infections, which in some cases have caused massive recalls of ground beef, eggs, spinach, and even peanut butter.

SPOILAGE BACTERIA

Bacteria that spoil food can be considered the "miscreants of rot." While often disgusting, rancid food is actually the work of a benign group of bacteria. Rotting meat is far more capable than other foods of harboring a menagerie of malodorous compounds, because the proteins present in meat are themselves capable of being broken down into mixtures reminiscent of eggs, fish, skunks, and more.

Although a roast like this beef navel should come up to room temperature before it's cooked, leaving it in the "danger zone" — between 40°F and 140°F — for more than four hours creates a breeding ground for pathogens.

The "smell test" is a good method of determining rancidity created by spoilage bacteria. Consuming the bacteria, and the resultant rank food, may trigger some nausea, but the bacteria responsible for rancidity will not cause any extended periods of illness. Nonetheless, the conditions that promoted the development of spoilage bacteria are identical to those needed by other, more malignant strains of bacteria. Therefore, the main concern with rancid food is the potential presence of dangerous pathogens in addition to the identifiable actions of spoilage bacteria.

INFECTIOUS BACTERIA

Bacteria that infect food are either invasive or noninvasive. The two most famous, *E. coli* and salmonella, are both invasive. Once consumed, invasive bacteria penetrate the walls of our digestive system and begin to wreak havoc on our system. Most of these bacteria have both benign and malignant strains. Symbiotic forms of *E. coli* exist in our digestive tract; without them, we would fail to absorb many of the necessary nutrients from the food we eat.

Most types of infectious bacteria produce dangerous toxins that cause disease or illness. After ingestion, the bacteria feverishly multiply, quickly increasing their population inside our digestive tract. Before the infection is evident, the increase in concentration of bacteria in our feces dramatically raises the likelihood of transmission to another host. Soon the bacteria begin excreting into our systems a toxin that is the actual cause of the illness. Signs of infection from invasive bacteria will most often appear before signs from infection from noninvasive forms because of the damage the invasive bacteria are doing to our tissues in combination with the toxins being produced.

POISONOUS BACTERIA

Often the onset of illness from ingesting food previously poisoned by bacteria will be quicker than that of illness caused by invasive or noninvasive species. Unlike infectious bacteria, poisonous bacteria have already produced the toxins required for illness or disease. Active bacteria rapidly excrete toxins in the food. These may sometimes leave a sour or off-putting taste, but often the presence of toxins is unknown to the eater until it is too late, with symptoms manifesting soon after consumption. Raw meats and dairy are highly susceptible to poisonous bacteria because the environments in which they are processed have high transmissions of fecal matter.

In many cases, cooking can destroy poisonous bacteria. However, while the bacteria themselves will die off at certain temperatures, the toxins they have produced are left behind and ingesting those toxins inevitably results in an ill reaction. Therefore, you should discard without question any food that you suspect of having become poisoned.

EXPONENTIAL GROWTH OF BACTERIA

Bacteria reproduce by cell division, meaning that one cell splits and becomes two self-sustaining single-cell organisms, each of those two cells then split, and so on. Through this method, bacteria populations can grow frighteningly fast. Their growth is exponential — the population doubles with every cycle of reproduction.

Exponential growth is challenging to comprehend, but a good example is to track the growth of a single bacterium. The sequence of growth would be: 1, 2, 4, 8, 16, 32, 64, 128, 256, 512, 1,024, and after another 10 cycles 1,048,576. Thus, in just 20 growth cycles, 1 cell becomes more than 1 million cells. The rate of a growth cycle is based on a balance between temperature and pH levels (assuming that nutrients are available for the foodborne pathogens, which is usually the case). Depending on the type of bacterium and the balance between temperature and pH, a growth cycle could range from as long as 12 hours to as short as 30 minutes. A 30-minute growth cycle means that a single bacterium can produce more than 1 million bacteria in around 10 hours! In some cases as few as 500 bacteria are enough to cause infection, illustrating the priority of curbing growth in any way we can.

Methods abound that attempt to reduce the growth of bacterial colonies. Drying, curing, salting, and adding preservatives are measures that all aim to curb bacterial growth. High temperatures, especially those outside the danger zone, are also known to kill off bacteria; heat is therefore the basis of pasteurization and many techniques of sterilization.

> The most common foodborne viral pathogen, responsible for 58 percent of all foodborne illnesses in the United States, is the norovirus.

Viruses

As microorganisms, viruses are a hardy bunch. They are not considered to be alive, because they do not procreate or operate independently. Therefore, they do not die but rather have to be inactivated to prevent infection. Heat is the most common method of inactivating pathogenic viruses; cold temperatures do nothing. Un-like bacteria, viruses rely on living host cells, such as our own, to reproduce. Viruses attach to cells and use the interior functions of the cell to create duplicates of themselves. After copies are made, the host cell often dies and the copies of the virus flood the host and continue the cycle. Without living host cells to invade, viruses can lie dormant on food for extended periods. The population of the virus will never increase prior to consumption, but in many cases contamination with small populations of a virus is enough to produce illness after consumption.

COMMON FOODBORNE VIRUSES

The most common foodborne viral pathogen, responsible for 58 percent of all foodborne illnesses in the United States, is the norovirus, named after Norwalk, Ohio, the location of an outbreak in 1968 that led to its discovery. Rotavirus and hepatitis are two other virus families that are a common cause of foodborne illnesses. Both norovirus and rotavirus have similar symptoms and cause gastroenteritis, more commonly known as the "stomach flu" or "stomach virus." The symptoms — vomiting and diarrhea — also happen to be the symptoms of many other foodborne illnesses, the similarity of which results in many misdiagnoses.

TRANSMISSION

Viruses are spread primarily through fecal-oral transmission, and in most cases from one infected individual to another. In a food service establishment a single infected employee can contaminate surfaces and foods throughout the workday, further increasing the localized cases and chances of a larger outbreak.

Other common culprits for passing a persistent virus are filter-feeding mollusks, like mussels and especially oysters, which are often eaten raw. Contamination occurs when the water surrounding the beds of growing mollusks becomes infected. Mollusks filter the water around them, absorbing nutrients but also viruses and other waterborne pathogens, harboring them until we consume the mollusks. Widespread cases of infection have resulted from ships discharging untreated sewage near oyster or mollusk farms. Other methods of transmission can include contaminated public bodies of water (e.g., pools, ponds, bathhouses), though any contact with infected individuals, especially those with poor hygiene, carries a risk of further transference.

Parasitic Worms

Foodborne pathogens are most commonly found on the surface of meat that has been contaminated. Parasitic worms are the exception: they burrow inside the tissue of animals. Like other foodborne pathogens they are either invasive (i.e., they penetrate the gut walls and travel elsewhere in the body) or noninvasive (i.e., they live in the gut, reproducing or causing other damage).

TRICHINELLA SPIRALIS

The most notorious foodborne parasite is *Trichinella spiralis*, a roundworm responsible for the disease trichinosis and the main concern for those processing land animals for food. The trichina worms proliferate through the ingestion of infected muscle tissue. The larvae burrow into the flesh of the host animal, forming cysts and causing initial infection. There they lie dormant, waiting for the animal to die, after which another animal consumes the infected flesh. Once consumed, the larvae mature in the intestines and grow into adult worms. The adult worms reproduce, forming more larvae that, again, burrow into the flesh and await death and transference.

Pigs are viewed as the main risk animal on a farm because of their omnivorous diet, but these days cases of trichinosis are exceedingly rare. Since the discovery of *T. spiralis*, farmers are taking the appropriate steps to rid pork of infection, and in the United States there are fewer than 10 reported cases of infection from commercial meat a year. In fact, most cases are from hunters, because the meat of many carnivorous game animals holds a high risk of infection. Even so, the reported cases are only in the dozens.

TAPEWORMS

Other worms are also capable of inflicting damage on livestock. Two tapeworms, *Taenia solium* and *Taenia saginata*, can infect pigs and cattle, respectively. These worms have a similar life cycle to that of *Trichinella spiralis*, with one major difference: humans are the only host in which these worms can mature and reproduce. Therefore, humans are also the only living beings that can spread the larvae, which is done largely by fecal transmission. Cattle and pigs that are raised around poorly treated human waste or given food that is handled by infected hosts with poor hygiene are the main hosts for the larvae. Cases in the United States are extremely rare, mainly because of food safety regulations.

Protists

Most people are familiar with certain forms of *protists*, single-cell organisms that are most often found in bodies of water. Algae, molds, fungi, and amoebas are all part of the protist family. Rivers, watering holes, swimming pools, and the like are all teeming with protists, most of which pose no risk to human health. Pathogenic protists, on the other hand, pose high risks to our well-being. They account for only 2 percent of documented foodborne illnesses in the United States, but the death rate from protist-related illnesses is staggering — protists cause almost 25 percent of foodborne related deaths. (Compare that to salmonella, which infects more than 1 million people a year, around 11 percent of the cases of foodborne illnesses, and results in a similar number of deaths.)

AVOIDING CROSS-CONTAMINATION

Food, especially meat, is an inherently clean substance. By and large, the pathogens that we encounter through consumption are not harboring on food as a permanent home: they're merely squatting temporarily until they find residence in a human host. The squatting and the taking up of a new residence is called *cross-contamination*, the process by which one substance is soiled by another already-infected substance. A doctor sanitizes a scalpel not to keep it clean, but rather to prevent the introduction of new microorganisms. We do the same with our tools and surfaces when we process meat and should also adhere to a similar approach when dealing with food storage.

Cross-contamination occurs constantly. Everything acts as a method of travel for microorganisms en route to better lands where they can proliferate at will. For some, humans are the greener pasture, and the microorganisms arrive through pathways that may involve doorknobs, toilets, utensils, or simply being around another human. When we handle or process food, cross-contamination is a serious concern.

Certain substances are more likely to contain specific pathogens than others. An easy example is poultry, which should always have a dedicated cutting surface. The basic reason for this is that poultry meat has a high risk of containing one of many strains of salmonella. Consequently, we take steps to prevent cross-contamination by processing different foods on separate surfaces.

All food holds a similar risk of contamination. Fecal matter is the main culprit, and any surface is capable of spreading it. For this reason, we need to take safety measures during the processing of any type of meat. This includes carcasses, even of the same species. You may start the day breaking down a beef forequarter. When you're done and you plan on breaking down the hindquarter of the same animal, it behooves you to scrape and sanitize the surface prior to making that first cut. The same applies to switching between any home and kitchen processing tasks, whether they involve butchering different species or even chopping vegetables.

TOXOPLASMA GONDII

There are a host of pathogenic foodborne protists, but the substantial death rate is primarily due to the lethality of a single entity: *Toxoplasma gondii*. The U.S. Centers for Disease Control estimates that around 60 million people in the United States carry the *T. gondii* parasite. That is a shocking number, especially when you consider the low numbers of reported illnesses from the parasite. The reason for this low ratio of infection to illness is that the average immune system is capable of keeping the parasite at bay. Those who exhibit signs of toxoplasmosis (the disease caused by the parasite) are often misdiagnosed, because the symptoms are like those of the flu and may resolve themselves before the person seeks medical help. People with compromised or weak immune systems — infants, children, pregnant women, and those with immunity disorders — are more susceptible to illness, and potentially death, from toxoplasmosis.

PROTIST LIFE CYCLE

The life cycle of protists, including their proliferation, depends on fecal-oral transmission. Water, the most commonly contaminated substance, allows for the parasitic eggs, called oocysts, to travel and find a host. Once ingested, the oocysts mature into parasites in the gut of the host animal, causing illness and disease while also producing more oocysts that can be passed through the feces of the host, continuing the cycle. Humans can contract protist infections through ingesting oocysts in bodies of water, on poorly washed vegetables, or in the meat of infected animals that contain larvae in cysts. Notable protist pathogens that spread through these methods are *Giardia lamblia* (the cause of giardiasis), *Cyclospora cayetanensis*, and *Cryptosporidium parvum*.

Toxoplasma gondii follows some of these laws of transference, but its life cycle has some bizarre twists. The definitive host of the parasite — meaning the animal in which the parasite can mature and reproduce — is the cat, either domestic or wild. Inside cats, the *T. gondii* larvae mature and produce more oocysts that are spread through the cat's feces. Cats contract the parasite from ingesting the meat of animals, primarily rodents, that are infected with oocysts and larvae-cotaining cysts. Because the feces of cats are the primary method of transference, cat owners, and the animals on farms with cats, face a higher risk of infection. Simply changing the litter box can spread the parasite; herbivores eating grass that was contaminated by a resident (or feral or wild) cat may result in contaminated meat, and the contamination is then passed to anyone who consumes it.

PREVENTING TRANSMISSION

Heat is the primary method of killing the oocysts and larvae of protists. Oocysts from some of the more rare pathogenic protists are exceedingly difficult to destroy at temperatures less than 200°F — far beyond the temperature you'd want to cook your meat to. Fortunately, the most common protist, *T. gondii*, can be destroyed at acceptable cooking temperatures. A link to the guidelines for cooking temperatures to kill off protists can be found on page 329.

Prions

The most obscure foodborne pathogen is the prion. Many aspects of its origins and methods of transmission are unknown, despite extensive testing and research. What we do know is that the structure of a prion is incredibly simple: it's a misshapen protein. And we also know that when this pathogenic protein gets into a host body, it causes other prion proteins within the nervous system and brain to fold abnormally, leading to neurodegenerative disorders. The most famous prion is that which causes bovine spongiform encephalopathy (BSE), commonly known as mad cow disease.

How to destroy prions is another mystery. So far, all of the methods that have been tried — including incineration, chemicals, ultraviolet radiation, and ionizing radiation — seem to be ineffective. Thus, there is little we can do as meat processors to prevent prion transference. Any animals that show signs of neurological disease must be reported to local authorities, so you may consider testing for BSE in non-ambulatory animals older than 30 months that are destined for a compost heap rather than a slaughterhouse in order to prevent potential transfer. From a food safety perspective, few other measures help. Sanitizing is futile. Fortunately, the spread of BSE is rare in animals and any connection to human diseases is even more rare.

Preventing the Spread of Pathogens

KNOWING WHICH PATHOGENS are of concern is an important step toward reducing the spread of foodborne illness, but memorizing scientific names and symptoms will do nothing to stop the spread of pathogens unless you take the proper precautions each and every time you handle and process food. Remember that food is largely uncontaminated until it comes into human contact, and that there are steps you can take to limit contamination.

Personal Hygiene

The first and foremost step to preventing the spread of pathogens is good personal hygiene. Almost all of the 31 foodborne pathogens identified to date can be spread through fecal-oral transmission. Although repulsive to consider, the truth is that we're constantly consuming food that's laced with traces of feces, mostly from other humans. As soon as you accept that reality, though, you'll realize that proper hygiene is the key to food safety.

REGULAR, THOROUGH HAND WASHING

Good hygiene starts with clean hands. Washing your hands needs to be a regular activity throughout any day of processing food. It may sound surprising, but many people don't know how to properly wash their hands. A quick rinse and a splat of soap won't cut it. Alcohol-based and other hand-sanitizing solutions will work against bacteria, but their efficacy against viruses is low, so don't rely on them exclusively.

There is no quick fix; the proper washing technique for stemming the spread of pathogens involves time and attention. This is how you properly wash your hands:

- Start with hot water, not blistering hot but rather around 110°F, or as hot as possible while still allowing a regular flow over the hands.
- Use an antibacterial soap. Look for one without environmentally damaging ingredients like triclosan.
- Spread the soap between both hands, creating a lather.
- Scrub the front and back of the hands, as well as along the thumbs, between the fingers, and over the wrists. Spend 15 to 20 seconds doing this.
- Use a nail brush to clean under nails, especially at the beginning of any processing and after exposure to more concentrated contaminants like manure, hide, and digestive waste.
- Rinse thoroughly with hot, clean water.
- Dry with a clean towel, either paper or fabric.

Hand washing must take place *prior* to processing food, but each of the following situations should be *followed* with a thorough hand washing:

- Soiling hands (e.g., with manure, intestinal fluids, urine, dirt, hair).
- Handling waste and garbage (e.g., inedibles, entrails, hides).
- Going to the bathroom.
- Blowing your nose; sneezing or coughing into your hand.
- Switching species of meat or carcasses.
- Handling chemicals (e.g., sanitizers, cleaners, disinfectants).

Hands that have open wounds should be cleaned thoroughly, bandaged, and then covered with latex or nitrile gloves. People who show signs of illness should not, under any conditions, be allowed to process food; the risk of contamination is far too great.

It is important to note that hand washing is not, in general, a method for killing pathogens. The soap acts as a surfactant, dislodging dirt, grime, and germs from the surface of your hands. After you thoroughly rinse them off, the pathogens go down the drain of your sink. Foot- or knee-controlled faucets and automatic soap dispensers, while optional, are helpful in ensuring that clean hands are not recontaminated while washing or rinsing.

The whole hand-washing process should take at least 30 seconds — far longer than most people typically spend scrubbing their hands. This may sound excessive, but it's really not. Strict hygiene policies will prevent potentially fatal illnesses from spreading while you're

Any sink that can be operated without hands, in this case with foot pedals, makes an ideal hand-washing station. An automatic soap dispenser would make it even better.

processing food. It's not enough to wash your hands properly at the beginning of processing and then briefly rinse them throughout the workday. Every time you wash, you should follow the procedure outlined above. Accidents do happen, but having a habit of good hygiene as standard policy will show that you are committed to creating clean food.

SANITARY CLOTHING

Washing your hands is certainly the most important step in achieving proper hygiene, but it doesn't end there. All clothes worn during the processing of food should begin clean and be kept sanitary. During slaughtering, it's helpful to wear a rubber apron that allows for repeated spraying and cleaning after any soiling from manure, blood, or any other contaminants. It's necessary to also wear an apron while butchering; have a few on hand that you can change out during the day. You should wear a hat to restrain hair. Your sleeves should be rolled up, and their edges kept clean. Fabric towels should begin clean and be replaced during the course of the day as they become soiled. Avoid placing towels on anything but sanitized surfaces.

Workspace Hygiene: Tools and Surfaces

All tools should be sanitized at the beginning of any processing, and should continue to be sanitized throughout the day as they become soiled. Dropping a tool, placing it on an unsanitary surface, and using the same tool on different carcasses are all ways that tools become soiled. Tools that penetrate the surface of meat, such as boning hooks and mechanical tenderizers, are especially prone to spreading pathogens because they introduce microbes to interior areas that are often not cooked to the same temperatures as the exterior.

SANITIZING

QUICK DIP. Professional processing environments often employ the quick-dip method of sanitizing. This frequently includes a vat of water kept between 180°F and 190°F. After thoroughly rinsing the tool under hot water with soap — dirt, grime, manure, and the like need to be initially removed — you dip it into the scalding water for a brief period, killing any potential pathogens on contact. The quick dip is an excellent method and can be easily replicated with a pot of water on a hot plate or, depending on the capability of the unit, with a slow cooker set to high. Ensure that the water is maintaining the correct temperature by testing it periodically with a reliable and accurate thermometer.

SANITIZING SOLUTION. Rinsing tools in a sink with soap and then dipping them in a sanitizing solution is another method. Sanitizing solutions can be specialized blends of chemicals or simply a mixture of bleach and water. Many chemical sanitizers require exposure for at least a few minutes; make sure you read through the directions for any commercial products to ensure proper usage.

Bleach and water is one easy and cheap solution for sanitizing everything from tools to surfaces. A mild mixture of one tablespoon of bleach per gallon of water will adequately sanitize any surface exposed to it for a minimum of two minutes. To properly sanitize tools, fully submerge them in the solution for the required two minutes. Surfaces should be liberally sprayed and left to sit. You don't need to rinse off this solution; in fact you shouldn't rinse, or you risk recontaminating the surface. Wipe surfaces with a clean towel and allow them to air-dry, after which the chlorine in the solution will evaporate.

There are many other household solutions that can also aid in disinfecting surfaces. Vinegar, alcohol, hydrogen peroxide, lemon juice, and salt are all commonly found in homemade cleaners. In general, they do work for clearing away some bacteria (some work better than others), though none of them is especially effective at disposing of viruses, especially the norovirus, which is the leading cause of foodborne illness. Because of this, they're not recommended as a solution for sanitizing food-preparation tools and surfaces.

One exception is the successive application of vinegar and hydrogen peroxide, which has been proven to be effective at killing just shy of 100 percent of surface pathogens on contact. Use a vinegar that is 5 percent acetic acid (regular white vinegar) and a 3 percent hydrogen peroxide solution (the basic drugstore concentration). Rather than mix the two, apply each one separately — the order doesn't matter — and then allow it to sit for a few seconds before wiping clean. This safe, effective, and cheap combination can be used for tools, surfaces, or food products like fruit.

Surfaces should be cleaned regularly throughout the workday, especially when switching between carcasses. Prior to sanitizing, use a bench scraper to clear away meat bits, residues, and fats, which inevitably accumulate during butchering. Further cleaning, with scouring pads and a surfactant, will help prepare heavily soiled surfaces, which should be relatively clean before sanitizing. Small cutting boards can be submerged for sanitization; large cutting surfaces should be sanitized by applying a liberal amount of cleaning solution to the entire surface and allowing it to sit for the appropriate amount of time.

> The successive application of vinegar and hydrogen peroxide has been proven to be effective at killing just shy of 100 percent of surface pathogens on contact.

A bench scraper is the ideal tool for keeping a cutting surface clear of debris and fat while you're working.

Cold Storage

SAFE FOOD STARTS with proper hygiene during slaughter and butchering, but the necessary precautions do not end there. Remember that all food is contaminated to some degree. Cold storage procedures, during every phase from slaughter to consumption, are equally important.

The one and only goal when storing meat or any other perishable product is to limit the growth of microbes. These include spoilage bacteria as well as the plethora of pathogens that may already be on the food but in benign numbers. The pathogens that are affected by cold temperatures are bacteria and parasitic worms. Bacteria are prolific at temperatures above 40°F; their multiplication rate increases with the temperature until the heat becomes high enough to kill them, which is usually around 140°F. This is the basis of the so-called danger zone: perishable foods left exposed in temperatures of 40°F to 140°F for extended periods will undoubtedly spoil and increase the risk of dangerous infection.

Refrigeration

Refrigeration and freezing are the most common methods of storing perishable foods. The moderately low temperature of refrigerators does not halt all bacterial activity; it merely slows down the growth of bacteria to buy us more time to consume it. As you probably know, food will certainly go bad in the fridge if you leave it there long enough. As you lower the temperature, bacterial activity continues to slow down until it basically halts, which is around 0°F, the operating temperatures of most home freezers. At these low temperatures food will not spoil from bacteria, but it will still go rancid. The bacteria do not die but are preserved, just as the food is, becoming active once again with the rising temperatures of thawing and cooking.

To be effective, all areas of a refrigerator — top shelves, bottom shelves, and drawers — need to be kept below 40°F. Many consumer refrigerators will show a mean temperature below 40°F when in fact some areas are well above that. The only way to be certain that a refrigerator is cooling sufficiently is to place several small glasses of water in various areas of the fridge, allow them 24 hours to stabilize in temperature, and take a reading from each glass, using an accurate digital thermometer. Adjust the fridge as necessary to ensure that the overall temperature is safe.

PROPER ORGANIZATION OF FOOD

Organization is also an important tool for eliminating some of the classic cases of cross-contamination in the refrigerator. Open containers of lettuce sitting beneath packages of raw fish or other meats, juices of thawing meats spilling out into vegetable drawers, marinades splashing on surfaces — these are all examples of how germs can easily spread in the refrigerator. All meats should be stored on the bottom shelf and usually covered. (There are, however, times when meats should be left uncovered and exposed to refrigerated air, such as when you want to dry out the skin of a whole chicken that is destined for roasting.) When defrosting frozen meats, place them in a container or on a plate to contain drips. Finally, avoid storing raw meats together with cooked or ready-to-eat products.

Freezers

Freezers are often as inaccurate as refrigerators are. Many home models will claim to maintain a 0°F temperature when in fact they fluctuate. The lower you keep the temperature, the better kept the contents will be while minimizing the effects of inevitable fluctuations. Stand-alone freezers are far more effective than those integrated with refrigerators, though they require space and additional costs. (For more on freezing, see chapter 8.)

> Perishable foods left exposed in temperatures of 40°F to 140°F for extended periods will undoubtedly spoil and increase the risk of dangerous infection.

Killing Pathogens with Heat

IT'S A RELIEF TO KNOW that most of the foodborne pathogens we're concerned with, except of course for those pesky prions, can be killed with heat. This doesn't mean that every piece of cooked meat is safe to eat, but if you follow the recommended guidelines for cooking, in most cases you'll reduce the chances of infection to a negligible number. To achieve the desired results, you must follow the guidelines for time and temperatures, and use an accurate digital thermometer to ensure safe cooking.

Cooking Whole Cuts

It's relatively easy to gauge the cooking time for a whole, unpunctured muscle, because the interior can be considered sterile. Such muscles include roasts, steaks, shanks, and any other cuts in which the only area of the meat that has been exposed is the exterior surface. The surface of the meat can therefore be held at a certain temperature for a given time to kill off any residing microbes. In some instances, this may be far less time than what it takes to cook: exposure to temperatures over 160°F will destroy or inactivate pathogens in less than a second. The key is to ensure that the actual surface of the meat is achieving the target temperature.

When searing on the stovetop, this won't be such an issue — any browning done in a pan occurs at temperatures that immediately dispatch microbes — but oven-roasting can be more complicated. It's not enough to set the oven to 350°F and consider it a done deal, because ambient heat does not equal surface temperature. For oven-roasting, a digital probe thermometer comes in handy. Every so often, stick it just under the surface of the meat and take a reading until you are certain you are hitting the target temperature. And remember to sanitize the thermometer before each use since you are puncturing the meat.

More Susceptible Products

Meats that have been ground, mechanically tenderized, or injected with marinades or brines — along with homegrown pork, wild game, and other meats susceptible to internal infection by parasitic worms — need to be handled

The thorough mixing of surface contaminants and pathogens makes ground meats especially risky for foodborne illnesses.

differently than whole, intact muscles do. The interior of such meats is just as likely to be infected as the exterior is. Hence, we need to heat all parts of these meats to temperatures high enough to kill any contaminants.

For example, the exterior of a steak needs to be exposed to a temperature of 145°F for a minimum of three minutes. If the meat you're cooking is a hamburger, the interior of that burger needs to be held at 145°F for a minimum of three minutes. If you prefer your hamburger medium-rare, which corresponds to a temperature of about 135°F — the U.S. Food and Drug Administration and the U.S. Department of Agriculture will claim that 145°F is a medium-rare target temperature, but if you follow that guideline you will quickly realize that the temperature is too high — then it's recommended to hold that internal temperature for 36 minutes. Without some fancy modern cooking equipment, this is impossible; it is for this reason that you assume the risk of eating so-called undercooked meats. We've all done it, and it's not to say we won't again, but it is good to understand what the risks are before we assume them.

For more information on proper cooking temperatures for various cuts of beef, see the Safe Minimum Cooking Temperature link on page 326.

TOOLS & EQUIPMENT

TO EXCEL AT BUTCHERING, you need only a few basic, modestly priced tools. This chapter covers those tools and the other equipment required for nearly every butchering task described in this book. The main sections include an alphabetical list of the basic tools, an explanation of what makes each tool ideal for butchering, and advice on which to choose. Further on, the chapter covers additional tools and equipment, some task-specific, that allow for more customized or advanced preparations. You may keep these specialized tools around for occasional use, but they won't end up sitting on your block like the everyday ones will.

Knives

KNIVES ARE AVAILABLE IN MANY DESIGNS and are marketed under many names, so choosing the right knife can be a daunting task. When beginning to shop for a knife, do so in person in a store that will allow you to handle the knives prior to purchase. The knife you choose should feel good in your hand since you will be gripping it tightly for long periods. Test it out using both of the grips described on page 50. Make sure it is easy to grip and feels balanced when you hold it; also make sure that the handle is made of a material that is easily sanitized. Wooden handles often benefit from a quick sanding with a high-grit sandpaper to remove finishes and provide a textured surface for easier handling, especially because they inevitably get covered with fat and other slippery residues. All knife blades should be made from a high-carbon stainless steel, which resists rust and stain while maintaining a sharp edge.

FLEXIBILITY

FLEX BLADES Allow for more nuanced cutting; great for shaping, denuding, and seaming; used mostly by experienced butchers.

SEMI-STIFF (semi-flex) blades Allow for enough bend to keep the edge close to a bone or table, especially helpful when boning a loin, separating a top blade steak from the scapula, etc. Great for jointing.

STIFF BLADES Allow for the most efficient transfer of energy into cutting. Precise overall; no risk of wandering when trying to make straight cuts. Great for jointing.

SHAPE

CURVED BLADES Best for accessing the small spaces in joints or hiding.

STRAIGHT BLADES Better for precision cutting, trimming, and portioning.

Blade Characteristics

Knife blades are available as flex, semi-stiff (sometimes known as semi-flex), or stiff. Semi-stiff and stiff blades are the two most common options for butchery and are best for those beginning to learn the nuances of cutting, following the curvature of bones, and making their way through joints. Flex blades, with their tendency to wander and change angles with less force, require deft handwork and are more often found on the blocks of experienced butchers. Semi-stiff blades allow for enough bend to keep the edge close to a bone or table, especially helpful when performing tasks like boning a loin or separating a top blade steak from the scapula. Stiff blades allow for the most efficient transfer of effort and energy into cutting, as none is lost in blade flexibility. The lack of flex also makes them more precise overall; there is no risk of wandering when trying to make straight cuts. Both are great for jointing, though the small amount of play in the semi-stiff blade can be forgiving when trying to access tight spaces.

Most knives are available in two shapes: straight and curved. Both shapes perform well, so the decision of which to use is going to be largely one of personal preference. Straight knives can feel like a more natural extension of your hand and tend to be better for precision cutting. Curved knives have a spine that slopes away from the edge. The point of the curved blade is also accentuated and designed to allow for access to small spaces, helpful for tasks like jointing a chicken or separating vertebrae. Furthermore, a curved blade works well with incremental cutting actions, such as removing beef hide or denuding connective tissue from curved surfaces.

The most common combination is a semi-stiff curved knife for boning and several stiff straight knives for trimming and cutting, and I would certainly encourage you to start there.

Types of Knives

Knives come in myriad shapes and sizes, with specialized designs intended to improve performance for various tasks. Having a small arsenal of knives at your disposal is fun, but for butchering you actually need only two or three of the right knives, and this is where I would start.

BONING KNIFE

Thin-bladed boning knives range from 5 to 7 inches. The boning knife will be your main knife, used for everything from separating primals, subprimals, and the like to boning, jointing, and trimming connective tissue. Start with

a smaller blade length and, once you're comfortable with the relationship between your knife and non-knife hands, consider using a longer blade.

BREAKING KNIFE

A breaking knife is any long-bladed knife that's used to break down primals into subprimals and beyond, and to perform many trimming tasks. Make sure it's 8 to 10 inches long, with a stiff blade that is either straight or slightly curved. A breaking knife may be found under many other names, including *butcher knife* and *sticking knife*; find one based on size and shape rather than name.

BUTCHER KNIFE OR CIMETER

The term *butcher knife* tends to be a catchall for any long-bladed knife, but in essence a butcher knife is around 12 inches long and has a straight, stiff blade. The main function of this knife is to slice cleanly through large cuts of meat, leaving no marks from dragging blades or sawing motions, for example, slicing boneless steaks or completing the separation of a beef round and sirloin after the bone has been split. It will be the knife that needs sharpening the least since it will not come into contact with things like bones, joints, and the toughest connective tissues. Keep it sharp so that your cuts are as precise as possible. An alternative option for a butcher knife is a cimeter (also spelled scimitar), which, although its shape is slightly different, can perform all the same tasks.

Knife Accessories
HONING ROD/KNIFE STEEL

The primary objective of knife care is keeping the edge sharp; often that has to do more with honing (realigning microserrations in the blade) than with actual sharpening (removal of metal from the blade edge). The rods on the market are made from many different materials: polished steel, grooved steel, ceramic, borosilicate (Pyrex), and steel with an embedded diamond surface. The only one that won't remove material from your edge is the polished steel. Grooved rods, while wildly popular, can cause deformation of microserrations on the blade edge rather than promoting alignment. Ceramic, borosilicate, and diamond rods all hone while, due to their extreme hardness, also removing small amounts of material. Thus these materials allow for minor sharpening and upkeep along with honing.

For the price, a ceramic rod with a fine grit (i.e., a grit of 1,000 or greater) is the best bet, but for pure honing without the risk of edge damage through minor sharpening, a polished steel rod is ideal. You may also want to consider owning multiple rods in different materials: a polished steel for daily use and a fine- or coarse-grit rod for minor upkeep. Look for a rod that is at least 10 inches long, choosing a longer one for easier use with your butcher knife or cimeter. If you plan on attaching it to your scabbard chain (described on page 40), make sure it has a metal ring on the end of the handle.

Bone saw

Aligning the edge of your blade while honing requires very little force.

SHARPENING

When honing no longer produces a sharp blade edge, it is time to sharpen your knife. Ask your favorite restaurant where they sharpen their knives or find a professional in your area who offers sharpening services to reputable culinary establishments. You can also sharpen knives at home with great results using diamond, oil, or water stones. (I recommend any of the sharpening systems made by Edge Pro. See page 329.)

SCABBARD

Working with multiple knives on a tabletop can be dangerous. Commonly during butchering a loose knife falls out of sight within a pile of meat, presenting a serious danger when you try to move the meat and find yourself gripping a blade. Furthermore, loose knives may fall on the floor, damaging blade edges while becoming contaminated. Using a scabbard is one easy way to keep your knives easily accessible while protecting yourself and those working around you. It will also help you keep your knives sharp and clean.

Scabbards, and the chains that hold them up, are available in two materials: plastic and aluminum. Plastic scabbards are a one-piece construction and hold up to three knives and a meat hook. They are heat-resistant and therefore make sanitizing easy when working with knife sterilization boxes. An aluminum scabbard is a multipiece construction that comes apart for easy cleaning and sanitizing. The soft metal helps prevent dulling of knife edges, and the large size allows for up to five knives. Either plastic or aluminum will do the job just fine, but choose the aluminum if you plan on regularly wanting more than three knives at your disposal.

Other Cutting Tools
CLEAVER

Cleavers are traditional butchery tools but are not used frequently in modern meat cutting. For most bones you'll pull out a handsaw or go to your bandsaw if you have one, because the results are cleaner than those from using a cleaver and don't carry the risk of

Sharpening with the Edge Pro ensures a consistent angle every time.

A scabbard protects your blades while helping to prevent accidental injuries from misplaced knives.

leaving bone shards in your product. But this doesn't mean that cleavers have no place. (One thing to consider is the cathartic effect of whacking your way through soft bones.)

Cleaver options run the full gamut of style, size, weight, and cost, but the most important factors are weight and the comfort in your hand. You want a cleaver that has heft, because its job has to do with impact, not precision. It should be top-heavy, like an ax, so that when you swing it the weight of its broad blade, not your exertion, creates the momentum. In contrast to many other tools of the trade, antique cleavers can be just as effective as new ones and often are more affordable for the weight.

BONE SAW/HANDSAW

The blades of bone saws, especially the teeth, are designed to easily and efficiently do one thing: cut through bone — not meat. The quality of cuts that can be made with a bone saw far exceeds those made by other saws you may have on hand — such as a hacksaw, a carpentry saw, or a bow saw. A strong stainless-steel or aluminum frame provides the necessary rigidity; the thin profile enables splitting within difficult spaces, like a beef sternum, and also prevents obstruction. The saw handle should be made from a material that is easily sanitized while being comfortable to grip. Blades should also be made from stainless steel and, when loaded, have a tension tight enough to prevent wandering while cutting.

Your choice of saw length should be based on the largest animal you will be working with. You should have enough room to freely move the saw in large strokes without the tip or handle having to enter the space you're splitting. Trying to split a beef round from the sirloin using a 17-inch saw would be difficult, because the strokes would be short and the blade length would not fully traverse the area that needs to be split. For that job, use a 25-inch saw or larger, depending on what is also comfortable to hold and manage when sawing. Further details on using a saw can be found on page 51.

Tabletop Equipment

IN ADDITION TO CHOOSING equipment that slices and severs, a butcher needs to keep a selection of other tools intended to aid the job of cutting meat on a table or hook. Every item on this list is essential for my process, and I suspect you will find value in them as well.

Bench Scraper

Your work surface will inevitably get gummed up with fat residues, small meat scraps, bone dust, and other butchery debris. Keeping your workspace clean is part of conscientious meat cutting, and using a bench scraper regularly while working is the best way to clear your surface of such detritus. This tool works equally well with large butcher blocks or small cutting boards. Find one made of metal with a comfortable handle that can be sanitized easily.

Bone Dust Scraper

Anytime you saw through a bone-in cut of meat — such as bone-in steaks or crosscut beef shanks, or lamb leg steaks — some debris is left on the cut face. This is especially evident when working with a bandsaw, which usually combines the bone dust with fat and moisture to form an unpalatable pasty mess. While you can use the edge of your knife to wipe the dust away, the lateral movement is not good for edge sharpness, especially around bones, and you run the risk of damaging the cleanly cut face of meat. It is better to use a bone dust scraper, a simple, cheap tool that does an effective job, leaving the face of your meat presentation-ready. Scrapers come in either plastic or metal, and both are equally effective.

Boning Hook

A boning hook not only is an essential tool for breaking down a carcass on the hook but also provides great benefits for table butchery as well. Its sharp, precise tip makes pulling and handling pieces of meat easier than gripping the carcass, and the length of the hook positions the non-knife hand away from apparent danger. A hook can also be used to help clean bones when dealing with marrow or when frenching. Find one with a comfortable handle and a hook length of at least 5 inches. Dulled or bent tips can be sharpened with a rasp or a grinder.

Butcher's Twine

Butcher's twine is a versatile tool; having some around, not just for tying off roasts and trussing chickens, will be helpful. Twines come in various ply ratings — such as 16-,

A bone dust scraper is essential for removing the unpalatable debris left after portioning a bone-in cut.

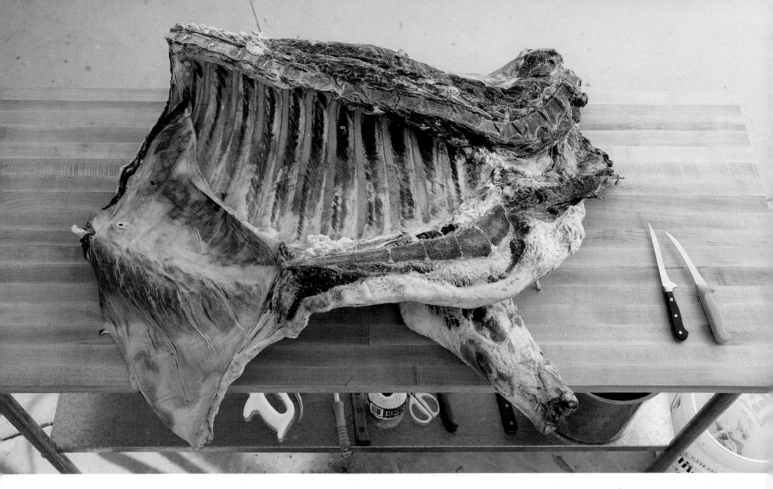

It's important to choose a work surface that's big enough to accommodate the largest carcass you plan on butchering.

24-, or 30-ply — which indicate the number of strands twisted together to form the twine; the higher the number, the stronger and thicker the twine. Get 100 percent cotton, food-grade twine that is 16- to 24-ply.

Cutting Board/Surface Material

There are two main requirements for a good work surface: it should be easy to sanitize, and it shouldn't cause damage to your knife edge. The two main options are wood or plastic (polyethylene) and, while the commercial industry tends to use plastic, wood is my preferred option, for several reasons. Wood is a self-healing material, meaning that over time as you make shallow cuts and grooves, it expands to fill the space, leaving fewer crevices for bacteria to reside in and develop. Dry wood also has natural antimicrobial properties due to the capillary action of its fibers, which absorb surface bacteria and kill them. Plastic, in contrast, is easily damaged from knives and saws, and the damage is permanent. Repeated use leaves deep grooves that are difficult to sanitize, even with soaking. Wood is also repairable: you can sand down the surface time and time again (until it is too thin to work on). Plastic, once badly damaged, needs to be thrown away and replaced. So, while both materials can be sanitized using the same solutions, the efficacy of sanitizing plastic is largely based on the effects of previous usage.

KINDS AND CUTS OF WOOD

Woods, the ones ideal for butcher blocks, are also gentler on knife edges than plastic. The softer the wood, the easier it is on your knife (though the downside is a shorter life through more wear). A balance between softness and durability is what makes maple the most popular material for butcher blocks.

Wooden boards are made from three different cuts: end-grain, edge-grain, and flat-grain. The terms are the same as with a piece of plank wood: *end-grain* is wood standing on end, *edge-grain* is the wood on its edge, and *flat-grain* is a board sitting flat.

END-GRAIN is an ideal option for knife care, because the fibers of the wood are aligned vertically, letting the knife edge fall in between them. Among the three types of cuts, it is the strongest, most resistant to wear, and longest lasting; fittingly, it is also the most expensive.

TABLETOP EQUIPMENT 43

EDGE-GRAIN, the next-best option, is moderately durable, and although the fibers run horizontally, they are vertically stacked and good for knives.

FLAT-GRAIN will not work for butchering use, because its durability is low. Just envision standing on a plank of wood laid flat: it bends, compared to when it is on its side or end. Flat-grain is the cheapest option, and for good reason: it will need to be replaced regularly.

A large wood butcher-block tabletop is the best surface to work on, but for many folks portable cutting boards are the only option. Choose a board that is as large as possible for the working space you have and for the manageability of moving and cleaning. Or combine multiple boards to make a mobile tabletop that is easy to clean. End-grain boards should be at least 2½ inches thick, and edge-grain boards 1½ inches thick. The thicker they are, the longer they'll last and the better they'll resist warping (though with proper maintenance no board should warp). Avoid culinary cutting boards that have a juice groove on the outer edge, because the groove will prevent easily scraping the board clean and will inevitably collect debris.

Kitchen Scale

A kitchen scale is an indispensable piece of equipment when processing meat. Using a scale improves the outcome of many tasks: you can label freezer-bound packages of meat with the weight; you can weigh portioned cuts to allow for adjustments that ensure better accuracy; and you can follow recipes based on ratios and weights, like the ones in this book, with certainty. Make sure your scale is digital and provides a readout in both the metric system (grams) and the imperial system (ounces). It should have a capacity of at least 5 pounds (ideally more), be battery-powered, and be easily portable.

Spray Bottle

Keep a quality spray bottle around, filled with a sanitizing solution, to encourage frequent cleaning of surfaces and tools. (For recommendations of sanitizing solutions, see page 31.)

Towels

Towels are essential during processing. Keep a stack around, leaving some clean for tasks like drying hands and sanitized tools or surfaces, and some that can be soiled in tasks like wiping down saw blades and bone dust scrapers, and in intermittent clearing of surfaces.

Clothing

THE APPROPRIATE CLOTHING for butchering will not only keep you clean and protect you from injury but also improve the safety of the meat you cut by preventing cross-contamination.

Aprons and Butcher Coats

Protective clothing keeps you clean while reducing contamination of the product. Butcher coats, or frocks, are the most effective protective garments. Aprons can be worn on the outside of a coat and should be switched out once soiled. Aprons should also be changed when switching from working with poultry to working with other proteins. Choose clothing that fits snugly and reduces the amount of excess fabric that can easily get caught in machinery or soil.

Cut-Resistant Gloves

Wearing cut-resistant gloves on your non-knife hand can help you avoid minor cuts. Although you can still inflict a serious injury even while wearing gloves, they can reduce the damage. Gloves may be especially helpful when you're beginning to butcher and learning the relationship between your blade length and non-knife-hand positioning. You should change your gloves when they become heavily soiled or when you switch from working with poultry to working with any other protein or vice versa.

Cut-resistant gloves (left) can help protect your hands from injury. Nitrile gloves (right) are the best choice for covering hands that have cuts or open wounds.

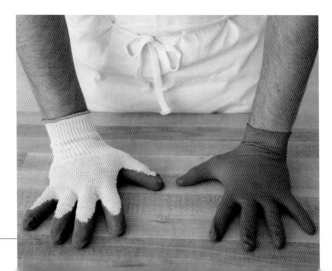

Nitrile Gloves

Always have a box of nitrile gloves on-site. If your hands have small nicks or any other kind of open wound, you should wear nitrile gloves to prevent contamination. Furthermore, wearing nitrile on the outside of cut-resistant gloves can reduce the need to swap the protective gloves when they become soiled or when you switch between different proteins. Choose a size that fits snugly but provides some room for the natural swelling of hands while working; they should never be tight at the wrist or feel like they're constricting blood flow to the hands.

Protective Aprons

Fast movements and sharp blades carry a high risk of injury. Some injuries may simply be nicks or shallow cuts, whereas others could be serious abdominal lacerations; you never know until it has already happened. One slip toward your stomach while knifing your way through cartilage could result in the equivalent of self-disembowelment. So, although wearing some protective gear is merely recommended, wearing something that protects your vital organs is essential.

There are three adequate options for abdominal protection, all of which come in the form of bibs or aprons: fabric, plastic, and chain mail. The fabric and chain mail exceed the protection of a plastic belly guard because they are able to cover more of the body without restricting motion, and it is certainly a benefit to have your groin and area blood vessels protected. The fabric from which butcher's bibs and aprons are made consists of lightweight cut-resistant materials, similar to what's found in cut-resistant gloves. Plastic is available in the form of a belly guard, which while not a proper apron wraps around the abdomen. Chain mail, or metal mesh, provides the best protection but is heavy and often expensive. Nonetheless, while 5 to 10 pounds (depending on the length of your apron) can certainly take its toll during a long day of butchering, the extra weight is well worth the protection. If you choose to use chain mail, consider wearing two fabric aprons, one underneath and one in front.

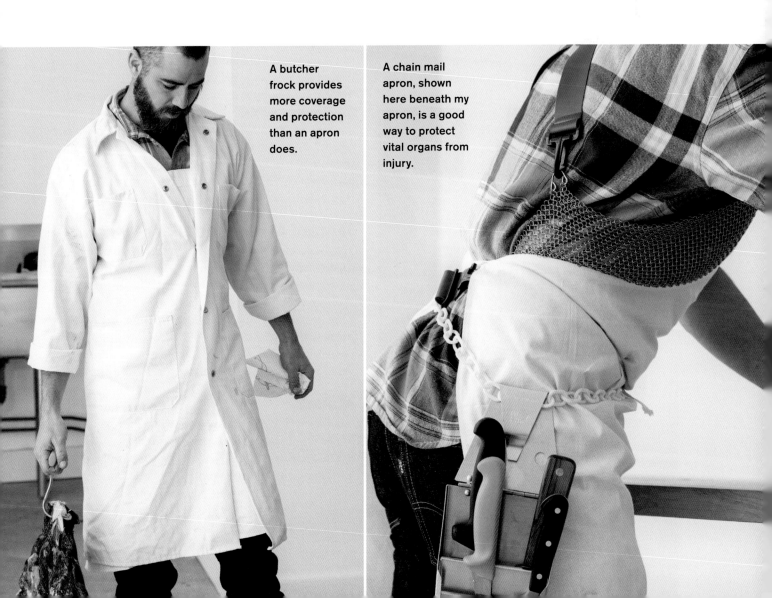

A butcher frock provides more coverage and protection than an apron does.

A chain mail apron, shown here beneath my apron, is a good way to protect vital organs from injury.

Optional Equipment

AS YOUR BUTCHERING SKILLS DEVELOP you may recognize a need for specialized pieces of equipment. Each of the following items provides a very specific function and is essential if you plan on doing things such as grinding or making cased sausages.

Bandsaw

A bandsaw is a powerful and versatile piece of equipment to have in a meat-processing shop. It makes quick work of many repetitive tasks like cutting steaks or crosscutting beef shanks, and can produce consistent results with the aid of measurement tools. The downside is that bandsaws are expensive, require heavy cleaning, and like all other machines need periodic upkeep. Everything that can be done with a bandsaw can, in theory, be done with a handsaw; the bandsaw may be quicker, cleaner, and easier, but it is not always better.

A bandsaw makes quick work of portioning a beef loin into even steaks.

Grinder

Ground meat is a common by-product of any whole-animal butchery, because it is one of the most effective ways to maximize carcass utilization, turning unpalatable bits into hamburgers, sausages, or other value-added products. Meat grinders come in all forms, from hand-crank models to kitchen mixer attachments to high-horsepower machines.

MANUAL GRINDERS are a challenge to operate for large quantities of meat; they're meant for the hobbyist.

GRINDER ATTACHMENTS for kitchen appliances provide adequate functionality for grinding several pounds of meat at a time, though their small capacity requires frequent cleaning for proper operation. They also require you to have another appliance around, so unless you plan on having a kitchen mixer dedicated to processing meat (which could be great for making many value-added products), this won't even be an option.

DEDICATED GRINDERS are built for the sole purpose of grinding meat and perform the best of the three options. They range from home models with low horsepower to commercial machines with integrated paddle mixers and other features.

A meat grinder should have swappable plates for adjusting the size of the grind. The moving parts should be made from a durable material and should be easily dismantled for cleaning. All-metal construction is ideal, not only for durability but also because a chilled metal grinder stays cold longer than a grinder of another material, which helps prevent fat smearing and poor grind performance. Protective features can prevent injury from clothes or hands getting caught in the auger.

Jaccard Meat Tenderizer

Tough cuts of meat can benefit from tenderization (more on page 64); one useful tool for this is a Jaccard meat tenderizer. It uses dozens of sharp, flat knives to penetrate the muscle, severing and shortening the fibers, giving the impression of more succulent meat by making it easier to bite through. Along with helping tenderize the muscle, the Jaccard's minor penetrations can also help distribute marinades by providing multiple access channels into the interior of the meat.

Mallet

A mallet, whether made from wood or metal, should have a good weight and be head-heavy. You can pound out cutlets and flatten medallions with the side of a cleaver, but having a mallet around will make the task much easier and cleaner. Some models have two sides: one flat and

The forty-eight sharp knives in a Jaccard help tenderize tough cuts.

one textured, the latter being for tenderization and the former for flattening. I recommend using a Jaccard for tenderizing and a mallet solely for flattening or striking. Find a mallet that feels comfortable in your hand and has a large surface area to the head so that your strikes won't have to be accurate to achieve good results.

You can even use a mallet with a cleaver, striking the spine of a cleaver when it's in place to avoid inaccurate swinging. In this case look for a wooden or rubber mallet.

Meat Lugs

For home butchering, you'll need multiple containers to hold everything from offal to trim to inedibles. While you can use any plastic or metal container, lugs designed specifically for the task of meat handling provide some readily apparent benefits: they are constructed from heavy-duty and stain-resistant plastic that resists warping and cracking from exposure to extreme temperatures, optional lids allow for stacking, and interiors are designed with sanitization and cleaning in mind. Lugs come in a variety of sizes to fit the volume of meat you're working with, so purchase the largest size that works for your processing.

Sausage Stuffer

A sausage stuffer is necessary only if you plan on making cased sausages (not just patties) or quickly stuffing ground meat into bags. Hand-operated stuffers, while requiring more effort, carry the benefit of not heating the sausage mix during the stuffing process, which helps maintain the texture and quality of the product. Find one that has a capacity of 3 to 5 pounds and has interchangeable tubes for different casings.

Choose between a grinder attachment (right) or a dedicated grinder (middle) when deciding to grind meat at home. A basic sausage stuffer (left) is another option for the home butchering table.

4

BUTCHERING METHODS

FAMILIARITY WITH MUSCLE STRUCTURES and the skeletal map, knowledge about food safety and storage, and a collection of well-chosen tools will start you off on the right foot, but the real foundation to butchery as a craft is the mastery of methods. Understanding the conformation of muscle groups, the grain direction, the interworking of a joint, the connection points for seams: this knowledge is gained only through hands-on repetition. There is no substitute. The more beef you break down, the better you'll become at boning the sirloin and denuding the top blade, as long as you aim to improve and take note of new details with each cut. A carcass is chock-full of cues that you will have to read to know where and how to cut. Every time you cut, you can learn something new; greater experience leads to a heightened awareness of the underlying intricacies, carrying you down the endless path of method and skill refinement.

We must all start somewhere, though, and this chapter provides the information you need for a foundational understanding of butchering methods, and then some. This is not beginner's knowledge per se; there is something to learn here for new and old butchers alike. These methods form the backbone of butchery. It is critical to have a solid grasp of the information in this chapter before you attempt any meat fabrication.

Foundations

EARLY BUTCHERING WAS, no doubt, less sophisticated than today's approach, which relies heavily on extensive research and analysis of carcasses. The basis of the craft, however — using a tool to segment a carcass into convenient edible parts — is still the functional foundation. Hence, before making the first cut, you must understand the basic foundations — skills that have been handed down through the centuries and are still practiced by modern butchers.

Holding a Boning Knife

There are two ways to hold a boning knife that you'll use during most of the butchering process: the pistol grip and the basic grip.

PISTOL GRIP

The pistol grip is the primary grip you'll use with the boning knife: invert the knife downward and hold the handle with a closed fist, the same way you would hold a gun (hence the name). The knife edge may face in either direction, depending on whether you're drawing the knife toward you or pushing it away from you; the former is the more common usage. The pistol grip allows you to hold on to the knife tightly, allowing the use of greater force to make your way through cartilage, fragile bone, ligaments, and other materials with resistance. Be especially careful when drawing the knife toward you: after cutting through some part of a carcass, it's not uncommon to accidentally jerk the knife toward your body, hence the need for protective aprons (see page 45). In most cases when this book suggests the use of a boning knife, this is the grip you should use.

BASIC GRIP

The basic grip is how most folks hold a knife when they pick it up: hand gripping the handle with the knife facing up. Any knife in your arsenal can be held this way: boning, butcher, cimeter, cleaver, you name it. With the boning knife, it's used for cuts requiring accuracy and finesse, like navigating the shape of vertebrae while boning a loin. To further increase control, extend your forefinger onto the spine of the knife. Switch to this grip when the pistol grip feels crude or prevents the right angle of approach.

Pistol grip

Basic grip

Positioning the blade in the saw so that the teeth are facing toward you will give you more control, especially with tasks requiring accuracy and finesse.

Finesse with a Bone Saw

The use of a boning saw may seem straightforward enough, since its form is familiar and in many instances its operation mimics the similar tools found in a garage: load the blade with the teeth facing away from the handle and saw with an action that couples force with outgoing strokes. This works for sawing large round bones such as femurs or arm bones, crosscutting vertebrae (separating shoulders and legs from loins), or dividing other rigid structures.

Finer, more fragile bones present a different kind of challenge, one that is overcome with strategy rather than brawn. Examples are sawing through the ribs of a veal rack or producing veal loin chops from a short loin. The use of force behind the teeth of the saw blade will disrupt these structures, making the bone edges jagged and the cuts sloppy. Furthermore, sawing through the more petite meat structures requires securing them to the table, most often with the opposite hand. Maintaining a steady hand is a challenge in itself when plunging a saw forward, teeth dragging across bone. The solution to these predicaments may not seem obvious at first but becomes clear with very little practice: reverse the saw blade, teeth facing the handle, and allow the weight of the saw to be the only force behind the teeth. You have much greater control when drawing the saw toward you than when pushing it away, and this will allow for the deft work required to make clean cuts with fragile bones.

However, don't just plop the entire weight of the saw onto the bone and start sawing, expecting fine results. Instead, hold the saw so that the teeth are just touching the bone — they're sharp enough to cut, as long as you don't get impatient and try to muscle your way through — and then maintain this level of contact as you make smooth strokes with the saw. Don't rush it: just let the saw do all the work of cutting; your work is to maintain light pressure.

TECHNIQUES 101

Honing Rod

The frequent use of a honing rod is one of the keys to maintaining knife sharpness and ensuring clean cuts. The honing rod realigns the misshapen edge of the blade; this requires only light pressure, just a bit more than touching, and only a few good passes to do the trick. If you find that using more pressure is required to garner results, then the knife needs to be sharpened, not honed. There are two approaches to using a honing rod: tabletop or freehand.

FIND YOUR ANGLE

Start by placing the rod tip down and perpendicular to table. (A towel underneath the tip can prevent the rod from slipping or forming divots in wood work surfaces.) Place the heel of the knife, nearest to the handle, at the top of the rod. Aim to hone an angle around 22 degrees. This is easiest to achieve by starting at 90 degrees, directly perpendicular to the rod. Split that in half, giving you 45 degrees, and then split that in half again, giving you about 22 degrees. (If you know the specific angle of your knife edge, estimate that instead.)

BUTCHERING METHODS

TECHNIQUES 101

TABLETOP METHOD

For beginning butchers, the tabletop method is a great way to start. The rod is secure on the table, eliminating one of the variables (rod movement); this allows you to focus on factors like angle and pressure.

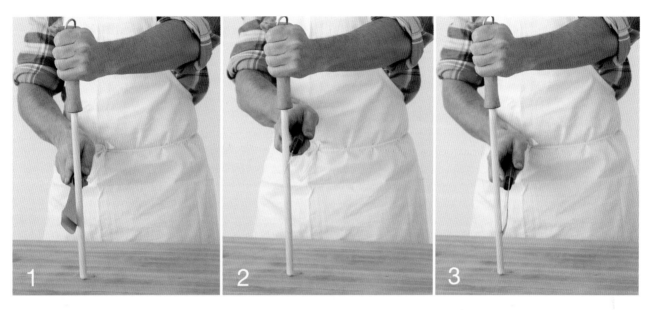

While maintaining your angle, draw the knife toward you while moving down the rod. Repeat with the opposite side.

FREEHAND USE

Once you've become comfortable with honing, integrating freehand use of the rod will be much easier than using the tabletop. This method uses the same angle and pressure, but the motion is performed while holding the rod in one hand.

Hold the rod, in the hand oppoosite your knife, at a slight upward angle. Place the heel of the knife, near the handle, at the tip of the rod, choosing an angle near 20 degrees. Draw the knife toward you, moving down the rod while curving your hand motion inward, ensuring that the entire edge makes contact with the rod. Maintain the angle and a light pressure throughout. Repeat with the opposite side of the knife, working on either the top or underside of the rod, moving the knife away from you (as shown) or drawing it toward you.

FOUNDATIONS

Trimming

AS A CARCASS GIVES WAY TO THE BLADE and muscles separate, falling from the bones, you'll come to realize that dissection is only the first step to producing palatable cuts of meat, a roughing out of the final product. Muscles are wrapped in fascial sheaths that connect with one another and converge to form tendon attachments to bone; an animal's glandular system is scattered throughout; halted suddenly, blood within the circulatory system coagulates as pockets in areas that went undrained; muscles require reshaping for culinary preparation. These, and many more, are all examples of how trimming is the true underpinning process of butchery.

Tenets of Trimming

A butcher is constantly trimming. It's the main action of cutting and is guided by a couple of ground rules.

TRIM AS LITTLE AS POSSIBLE

The first rule of trimming is to trim only what needs to be trimmed: nothing more, and certainly nothing less. Financially, trim represents lower value than whole muscle products. Ground meats and sausages are superb ways to garner value from edible trim, but, in general, meat is more lucrative when sold in a solid form, as portioned cuts or roasts. This is especially true when fabricating high-value primals like loins.

Trimming, like all other components of butchery, is a learned skill that gets refined through repetition and practice. When you first start butchering, you'll inevitably find yourself removing large swaths of meat during the trim process. This is okay. Focus on the priority, which is removing inedibles or shaping the cut mildly. (These priorities are further discussed on page 56.) Don't focus too much on the finer details of muscle curvature and final presentation. As your knife handling improves, your actions will become more precise. You will undoubtedly notice a natural reduction in the amount of meat you remove with each trimming motion, and it is then, and only then, that you should shift your focus to the next couple of ground rules. Otherwise it'll be like trying to cut a perfect circle out of a square piece of paper — you'll end up with a large amount of scrap paper (the trim pile) and a small circle (the piece of meat).

TRIM WITH THE GRAIN

When possible, trim in the direction of the muscle grain. Every muscle has a distinct although not always obvious grain direction. (Refer to page 6 for information about how muscle fibers create grain direction.) Think about those fibers as you would strands of hair. When you run your hand across the back of a dog, from head to tail, the hair lies flat and you can feel the fibers side by side. Switch directions and you have the canine equivalent of a bad hair day.

Muscle grain behaves in the same way. As a knife cuts meat, it severs individual muscle fibers. Cutting with the grain direction ensures that fibers remain tightly packed and maintain a uniform surface. Each fiber is sliced and

The grain direction of muscles is important to recognize, especially when you're trimming, so that the meat can maintain a smooth surface and attractive appearance.

TECHNIQUES 101

Trimming from Start to Finish

In the following sequence, a boned strip loin is trimmed for palatability and presentation: beneath the layer of fat that sits atop the loin muscle is a swath of silverskin that is best removed. First, we remove the excess fat, then the silverskin, trimming with the muscle grain the entire time. The excess fat can be trimmed and tied back onto the roast, a process called barding: the fat renders while cooking, basting the lean roast while adding flavor.

TRIMMING 55

laid next to another, in effect smoothing the surface. If the knife travels against the direction of muscle fibers (against the grain) it causes them to separate and lift, creating an uneven surface. This is most obvious with dense, thin-fibered muscles like the tenderloin, but it becomes evident in all muscles once you are aware of it. Additionally, the sharpness of your knife can dramatically affect the separation of the fibers. If you need to cut against the grain, use a recently honed knife and move slowly as you make the cut.

Removing Inedible Trim

The main priority when trimming is removing inedibles: the unpalatable pieces of meat and other bits found throughout the carcass. This may include anything from shriveled exterior surfaces to pus-filled glands.

ABSCESSES

Gland-looking growths that occur within a muscle are most likely abscesses and will need to be removed. An abscess is a sign of infection and, to be safe, a large area surrounding the abscess should be removed along with it. If you discover abscesses or other signs of infection or illness and are concerned about the safety of the meat, ask a local veterinarian to inspect the carcass.

BRUISES

During slaughtering, the process of stunning and the subsequent fall to the ground (along with some of the agitation that may lead up to slaughter) can result in bruised tissue. After an animal is skinned, the carcass should be inspected for bruising and any damaged tissue removed.

CONTAMINATED TISSUE

Between the time the animal was alive and the time its carcass landed on your butcher block, the meat you're cutting was exposed to contaminants. Areas close to where the hide was opened or the digestive tract removed are prone to contamination during slaughter. Furthermore, transportation of a carcass will often result in dirt or grime on the exterior, necessitating a careful inspection prior to aging or cutting.

DRIED TISSUE

Aging whole carcasses or extended storage in refrigerated spaces with ample airflow will cause surface drying of the meat. Discoloration and a hardened texture are both sure indicators of where to trim.

GLANDS

All animals have glands that need to be removed and discarded. Two glands that are exceptions to this rule are the thymus and pancreas, also known as sweetbreads. To be safe, if you are unable to identify a given gland, trim it away and dispose of it. All glands are located outside muscle, often among fat or connective tissue. The most common places to find large glands are at the convergence of large areas of the body — for example, at the leg and cavity, at the neck and shoulder. Smaller glands may show up in any other area of the body.

Separating Inedibles

Keep a separate container available for inedibles when you're butchering. Clearly differentiate it and keep it on the floor, on a pushcart, or on some other surface — never on the tabletop! — to help you avoid mistaking it for edible trim, especially when you're working in a team envi-

Dried tissue should be removed and discarded.

ronment. Useful ways of marking containers of inedibles include using clear labeling on the outside and lid; buying color-coded containers; or, if you are not composting or rendering, just using a common trash can. Never keep inedible trim on the work surface: immediately dispose of it. Tightly integrating this habit into your workflow will dramatically reduce the risk of contamination and spoilage. (A single sour-tasting gland can ruin an entire batch of ground meat.)

Removing Unpalatable Trim

Fascia, ligaments, and tendons are unpalatable, but this does not mean they are without purpose in other culinary pursuits. They can be used to strengthen stocks, create a consommé raft, and add body to sauces.

FASCIA

All muscles are covered by a continuous layer of connective tissue called fascia. It exists throughout the entire mammalian anatomy, providing structure and support to muscles, organs, and the body overall. In some instances its presence is negligible, while in others it takes the form of a thick, fibrous, impenetrable layer called *silverskin*. Because it's difficult to chew and digest, silverskin always needs to be removed — a process called denuding, or peeling, the muscle. (See page 58 for more detail.) Thinner membranes of fascia are often left untrimmed, since they may not detract from the tenderness of the meat or they may melt into gelatin during cooking.

TENDONS

The fascial covering of a muscle or muscles converges to form a tendon, the attachment to bone. Tendons, rarely consumed in the United States, should be removed and reserved for other uses. Since they are composed mainly of collagen, tendons can be added as a natural thickener to stocks, soups, or any slowly cooked moist-heat dish. Tendons are often confused with ligaments, which are also collagen bands; however, ligaments connect bone to bone; no muscle connection is involved.

Edible Trim from Shaping

Save the edible trim: its utilization is one of the tenets of a zero-waste approach to whole-carcass fabrication. Edible trim is the by-product of the second major trimming priority: shaping.

The natural form of muscles is not always advantageous to the methods we use to make them edible. The tenderloin is a classic example, with its elongated tail

Removing intercostal meat and fat caps is a common shaping task that produces edible trim.

that tapers as it extends from the butt end; another is a boneless sirloin, with its undulating shape after the pelvis is removed. These do not make for ideal shapes when aiming for cooking evenness or other culinary concerns. It is for this reason, among others, that we trim for shape. (Another way to affect shape is tying, discussed on page 70.)

Portioning cuts is a form of trimming for shape. It takes experience and skill to maximize portions from an oblong muscle. Cutting 1¼-inch-thick steak from a loin primal is one thing; lopping off 8-ounce boneless portions of top round is a far greater challenge. Portioning requires accurate estimation along with knowledge of the ideal shapes and orientations to maximize repetitive shapes (the overall yield) from an uneven object. Along the way, the process produces edible trim. (See page 65 for more on how to portion.)

Edible trim to be used for grinding should be placed in a clearly labeled container separate from those containing tendons and fascia. Excess fat should be cut away from the lean meat and kept in a third container, allowing for control over the fat content in ground meat or for use in rendering.

TECHNIQUES 101

Denuding

Denuding is the action of removing the exterior surface of a muscle or group of muscles. The word is most often applied to the removal of silverskin, and there are a couple of different methods. Here is an example of one method, showing how to denude a tenderloin while also demonstrating how to test for grain direction.

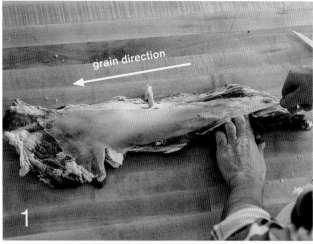

1. **BEGIN BY DETERMINING THE GRAIN DIRECTION.** Scrape the blade along the surface in both directions and notice which one provides resistance, maybe with some fibers lifting from the surface. Thinly trim a small area of the surface and run your fingers along it, feeling for fibers. The direction without resistance – in which scraping lifts no fibers and you feel no fibers when running your fingers along it – is the direction in which you want to cut.

On the top side of the tenderloin, this direction is from the tapered tail toward the butt end. The underside will always be the opposite, so in this case from the butt end toward the tail. Once you discover the grain direction for a muscle, it will be the same for other species. For example, pork, lamb, goat, and beef tenderloins all come from the same muscle group – *psoas major* and *psoas minor* – making the grain directions identical.

2. **CUT ALONG THE GRAIN.** Cutting direction follows grain direction: we now know to cut from the tail toward the butt end. Find the end of the silverskin and slide the knife underneath a portion of it. Secure the tenderloin by placing a hand on the muscle behind the knife. Angle your knife edge slightly toward the silverskin. Cut the length of the silverskin, moving your hand to stabilize the muscle as needed. Sever the strip of silverskin at its origin and then repeat until all the silverskin is removed.

>> **Some muscles — loin, sirloin flap, and clod heart, to name a few — are better denuded by removal of the entire silverskin in one piece rather than in stages, as with the tenderloin. This is often true for thick silverskins that cover large areas of a muscle.**

TECHNIQUES 101

Seam Cutting

Separating muscles along natural seams makes for more efficient cutting and maximizes the carcass yield. It also requires less effort, making for less fatigue. With a little force applied and the right connections severed, muscles tend to fall away from each other. Gravity may be all the force that is needed, though often it will require some muscle of your own.

The action, regardless of whether you are working on the hook or on a table, is basically the same. In both instances a boning hook is a useful tool for increasing the force applied with minimal exertion.

1. FIND THE SEAM. Start by finding a seam of fat and connective tissue that joins two muscles. This can be done at known junctions between muscles, as with this lamb shoulder, or by exploring with your hands, looking for areas where muscles meet. You can work your fingers into the seam to test for the ease of separation; some muscles, like those in the flank, can be separated mostly without the use of a knife.

2. SEVER ATTACHMENTS. Begin severing the attachments between muscles using a knife, staying in the seam. Keep the cuts shallow: you're mainly trying to peel the muscles away from one another, while the knife will sever the more stubborn connections.

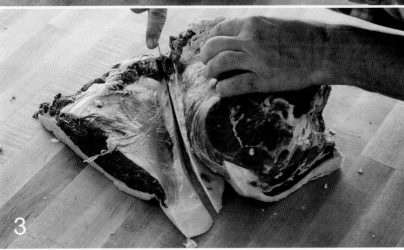

3. PULL AND SEVER. If you're working on the table, secure the underlying muscle with one hand while pulling with the other, peeling the muscles away from each other. If you're working on the hook, the underlying muscle will be attached to the hanging carcass; thus all your effort can go into pulling. Continue this back-and-forth action of severing attachments and pulling, keeping the cuts within the natural seam, until the two muscles are completely separated.

Boning

EACH BONE AND JOINT IN AN ANIMAL has its own characteristic shape, formed for specific tasks through generations of evolution. Understanding a bone's purpose can help you understand how that bone plays a role in the overall skeletal structure of an animal. And where there is a bone, there will be a joint. Every joint is formed by the intersection of uniquely shaped ends of two or more bones, often covered in cartilage and held together by ligaments and woven layers of connective tissue called *deep fascia*. Knowing the shape of these bones and the manner in which they fit together to form joints will vastly improve your ability to move quickly and efficiently through the process of boning beef necks, sirloins, loins, and so on.

Basic Boning

Boning is the act of removing one or more bones from a part of the carcass with the goals of maximum muscle removal and clean, separated bones. There is in ideal entry point for the removal of each bone and its associated muscle grouping. Through repetition (and some guidance in this book) you'll learn how to elegantly access a bone and remove it with minimal effort.

REVIEWING THE SKELETAL STRUCTURE

Before you bone an animal for the first time, take a quick look at a diagram of its skeletal structure. I cannot stress this enough. Five minutes with a diagram will save you time, effort, and money in wasted meat. Better yet, keep a copy nearby while you work. Take note of the specific bones in the area you're going to be working on. What is their shape? Where do the bones connect and form joints? What is their orientation? Asking these three simple questions will dramatically improve your results. If you're working on a leg and plan on removing the femur, you will know that it is a long bone with joints on both ends and that it runs from the hip to the knee. If you're boning a shoulder and have to remove the scapula, you'll know that it has a ridge on the side facing outward and that it is flat on the other side; it has only one joint and connects to the humerus in the arm. These are very quick observations that will inform you when deciding where to access bones and how to remove them.

Joints

Although it's possible to bone an entire carcass and leave the skeleton intact, chances are that won't be your approach. Most likely you'll be breaking down a carcass into primals and then boning, and to do so you will need to cut through some joints. You could do this using a handsaw, or even a bandsaw in some situations, but this would be missing the point. Learning the anatomy of an animal is paramount to butchering, without which the craft and

Before removing the neck, typically one of the harder bones to follow, it's important to understand its shape and orientation.

TECHNIQUES 101

Boning

Any boning job can roughly follow the same set of tasks. After some experience, your order of operations may shift here and there, but starting with this approach will help inform you before making adjustments.

In this example we're boning the humerus from a square-cut chuck. The humerus is a round bone, so we'll be working around a cylindrical shape rather than against a flat face. This means that cuts will need to be shallow to avoid cutting deeply into the underlying muscles.

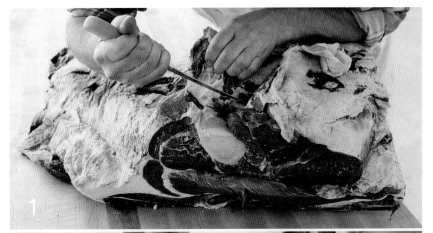

1. IDENTIFY THE BONE AND CHOOSE AN ENTRY POINT.
Familiarize yourself with the shape and location of the bone, along with any joints, then find a muscle seam that leads to the bone. Cut along the seam until you reach bone.

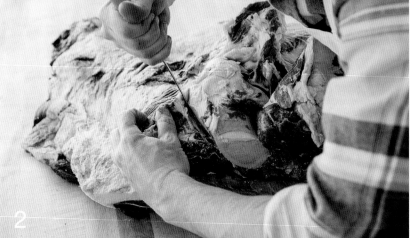

2. CUT ALONG THE BONE.
Upon reaching the bone, cut along the direction that it runs — not perpendicular to it. Move the blade along the bone in long, shallow strokes, releasing small portions of the muscle surrounding it and thus gaining access to the bone. Completely separate the top side of the bone, peeling the muscle away while using long strokes.

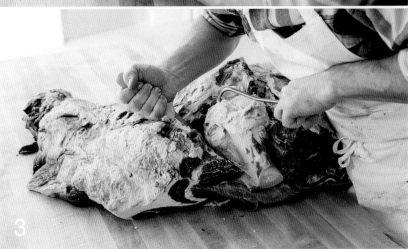

3. CUT UNDER THE BONE.
Find the edges of the bone, follow their shape, and start separating the underside. Stay close to the bone, separating one end before the other. In most cases, separate the joint-end of a bone as the last step.

beauty of being able to produce delectable cuts of meat with a simple tool (the knife) would be lost. Modern tools have their place, and can certainly make some tasks faster, but a solid understanding of anatomy is what will make you a great butcher. Mastering the joints is part of this.

A joint is the meeting place of two bones, held together by a combination of muscle, ligaments, tendons, or other connective tissue. Although you can look at a skeletal diagram to see the exact shape of a bone, learning the access points of a joint will take more trial and error — and quite a few sharp knives turned dull. To separate the bones, you'll need to understand the shape of the connecting parts (cartilage-covered bone ends) along with what connective tissue (ligament, tendon, etc.) needs to be severed to release the bond.

SEPARATING THE JOINT

When exploring how to separate a joint, go slowly. Don't try to cut your way into or through the joint with sheer force. Don't insert your knife tip into the joint and try to pry it apart. Not only will these approaches quickly dull or ruin your knife, but you're bound to cut yourself doing it. Instead, take the time to examine the area around the joint and note the obvious connective tissues. Make some exploratory cuts, use the tip of your knife to cut through some ligaments, and then apply force to the joint by moving the bones or trying to bend them over the table edge. Notice which incisions cause more instability than others do. Cutting through large tendons and ligaments is an ideal place to start. Don't be timid — you may at times feel like you're cutting through bone or cartilage because of the force it takes to get through them. After you knife through a joint a few times, the right method of approach will surely stay with you.

Peeling the Bone

Every bone is wrapped in a layer of fascia, which the musculature surrounding the bones attaches to. When boning, sometimes it's more beneficial to "peel" the bone — to get under the fascial layer and remove it with the muscle itself. Examples of such situations are removing the scapula from a chuck roast, separating the sirloin tip from a hanging carcass, and frenching the bones of a rib roast. These examples and others are explained with instructions in their respective chapters.

There are two different approaches to peeling the fascial layer off a bone, and they are easily differentiated.

REMOVING THE MUSCLE. With the first method, your goal is to cleanly remove a muscle, or group of muscles, that are attached to bone; the bone gets left on the carcass.

REMOVING THE BONE. The second method is the inverse: removing a bone while leaving a muscle or group of muscles still attached to the carcass.

Fortunately both methods employ the same skills. As mentioned earlier, butchering is largely a process of identifying and severing key connection points between muscles and bone and allowing parts to fall away according to natural seams. Peeling a bone is no different. Every bone has ideal approaches to releasing its fascial layer, and through practice you will understand how best to approach each unique situation.

Separating a joint is a combination of severing the right connections while applying force.

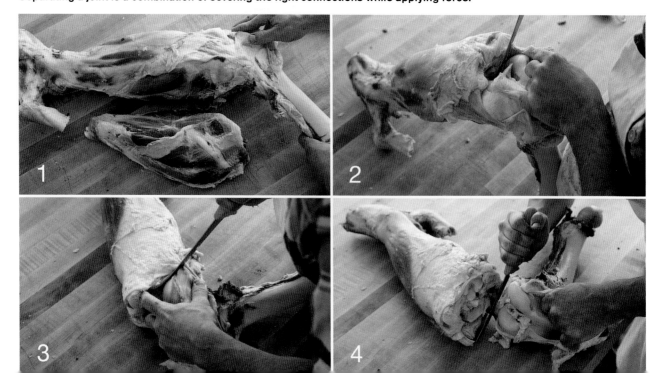

TECHNIQUES 101

Peeling the Bone

Here are the basics to get your explorations in peeling the bone started, using the sirloin tip as an example.

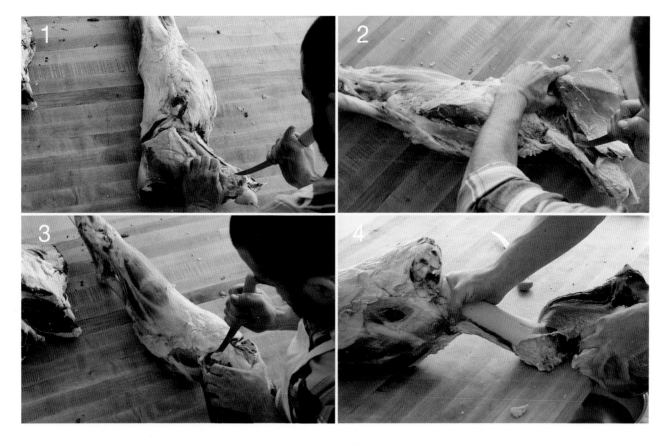

1. Find a muscle seam and make your way to the bone. On one side of the muscle, make a firm incision along the bone using the tip of your knife. (The first time you do this it might feel like you are cutting into bone itself, but you will soon realize that where you think there is only bone there is actually another layer of fascial covering.)

2. Repeat with the other side: find a seam, get to the bone, and then make a firm incision.

3. Sever the upper connection of the muscle, also ensuring that the cut goes deep to the bone and through the fascial covering.

4. Steady the bone with one hand. With a firm motion, pull the muscles toward you, peeling it away from the bone. Sever any remaining connections to finish separation.

MAKING USE OF BONES

Finding a use for bones will maximize your carcass yield. Any of the vertebrae or joints will be excellent stock bones due to heavy amounts of collagen and connective tissue, structures that break down under moist heat into unctuous gelatin. The round bones — femur, tibia, ulna/radius, humerus — can be crosscut for marrow bones, smoked for stocks, or used as soup bones. Ribs, scapulae, and pelvic bones are all great for dogs. By all means, find a use for the bones and get the most out of your animal.

Tenderization

RAW MEAT IS UNPALATABLE — our teeth struggle to sever the bundles of muscle fibers and make their way through the tough collagen. Raw meat that is served for consumption has been processed in a manner that overcomes this limitation and allows us to enjoy the experience: sliced exceedingly thin, as with carpaccio, or diced into small bits, as with tartare.

Cooking is the main form of meat tenderization. It unravels proteins; breaks down muscular structures; and, along with increasing the nutritional content, transforms a previously intimidating hunk of flesh into a delectable gastronomic experience. But cooking isn't the only tool we have to increase tenderness. We can also employ techniques to sever the fibers for us (mechanical tenderization) or use acids and enzymes to break down protein structures (chemical tenderization).

Cutting across the Grain

Discussed in the preceding section on trimming, muscle grain direction is an important trait to identify when butchering. Cutting with the grain is important for trimming, but if you employed the same approach when portioning, the results would be subpar. This is because the length of muscle fibers would be a challenge for us to chew through. Instead, we want to shorten the fibers whenever possible, thus increasing the tenderness of resultant portioned cuts. We do this by cutting the meat perpendicular to the grain direction. Kari Underly, a talented butcher and consultant for the meat industry, has named this technique the "90-degree rule," which helps illustrate the idea that you always want to cut exactly perpendicular to the linear direction of the muscle fibers.

Steak is one example of a cut that's produced by cutting across the grain. The main loin muscle, *longissimus dorsi*, has a grain direction that runs roughly parallel to the spine. Hence, steaks, whether bone-in or boneless, are cut perpendicular to the spine. Cutting across the grain becomes more challenging when the grain direction doesn't run so conveniently with the muscle itself, as illustrated in the photos below.

Mechanical Tenderization

Technically speaking, the 90-degree rule is a form of mechanical tenderization, since it employs a tool to do the work, but mechanical tenderization is more frequently applied to techniques that grind, pound, or sever muscle fibers into shorter forms of themselves. You've probably eaten a hamburger or sausage: in this form of mechanical tenderization the meat is ground into small pieces, aiding in mouthfeel while also creating a unique form. Meat grinders use manufactured dies, plates, and knives to break down characteristically fibrous muscles into

It can be difficult to cut a steak when the grain doesn't run parallel to the length of the muscle. In these cases, slice parallel with the grain to separate a section of the muscle, equal in thickness to your target portion. Then, slice perpendicular to the grain for final portioned cuts.

tender morsels. Mallets and meat hammers tenderize meat by separating the fibers, breaking apart the collagen structures that hold them tightly in bundles, and often creating flattened cuts like cutlets. The Jaccard is a tool that utilizes dozens of flat, sharp knives to penetrate meat and sever muscle fibers (see page 46). It serves the purpose of prechewing and can dramatically improve the texture of tough cuts. But be wary of overusing it, which can result in a mealy texture.

Chemical Tenderization

Chemical tenderization is more passive than mechanical tenderization. (Passive for us, though the action of the chemicals is anything but.) With this method, you expose the meat to a solution that aims to break down protein structures. Given adequate time, this will lead to dramatic improvements in tenderness; left to their own devices for too long, the ingredients in the solution will make the meat mushy and overtenderized. Common ingredients that aid in tenderization of muscle cuts fall into one of the following categories.

ACIDIC

The most common form of acidic tenderization is the traditional marinade, which often includes ingredients like citrus juice, wine, vinegar, or anything else with a pH below 5. The action of an acidic tenderizer is highly effective, breaking down both connective tissues and muscle proteins. An additional benefit is the increase in moisture-holding capacity, making the meat juicier as well as more tender.

ALKALINE

Though they are on the opposite end of the pH scale, alkaline marinades react with meat in a very similar manner to acidic mixtures. Frequent ingredients include baking soda, soda lime, lye, or anything else with a pH above 9.

ENZYMES

Fruit juices are a source of enzymes that snip proteins into segments, thus increasing tenderness. The following are the most popular fruits (with their enzyme names in parentheses): kiwi (actinidain), papaya (papain), and pineapple (bromelain). Bromelain is highly effective because it attacks both muscle and connective tissues. These enzymes can be found in liquid form (fruit juice) or in a concentrated powder, the latter being the common form found for sale in grocery stores. (Enzymes are discussed in more detail on page 18.)

FERMENTATION

The live cultures in fermented products can enact positive changes on meat, allowing the naturally occurring bacteria to attack muscle fibers, breaking them into fragments. Buttermilk and yogurt are two traditional fermented products used for tenderization. They do not break down collagen fibers, like other chemicals used for tenderization do, but they do provide an inhospitable environment for dangerous bacteria, thus retarding spoilage.

Portioning

MEAT MUSCLES CAN BE COOKED in any form, large or small, but there is a frequent need to cut a muscle into portions, be it individual steaks, a two-person roast, or sizes of specific weight. Portioning is a true skill, requiring repetition and time to maximize usable portions from amorphous lumps of meat. Some instances are easier than others — ripping 1-inch-thick steaks from a loin primal on a bandsaw is more a redundant activity than one of dutiful accuracy — but each action of fabricating portioned cuts should embody the same priority: maximize yield, minimize loss.

Portioning is based on either weight or size. Starting with an entire boneless loin, you can create a set of 8-ounce boneless steaks by weighing the entire boneless loin and then dividing the total weight by your target portion weight (in this case 8 ounces). You could also portion the loin by size, in this case thickness, by cutting 1-inch steaks. When portion weights are not critical, use the easier method of portioning by size.

Portioning muscles of inconsistent and oblong shapes (which, frankly, most muscles are) poses some other challenges to maximizing your yield. Consider this: when portioning for weight, the width for a portion will also affect the thickness. An 8-ounce portion that is 3 inches wide will be much thicker than one that is 5 inches wide. If the muscle is 9 inches wide, choose a width that divides cleanly, like 3 inches or 4½ inches.

TECHNIQUES 101

Portioning Based on Weight

Muscle of consistent and elongated shape, like this trimmed portion of a strip loin, is good to use as you start practicing portioning. In this case we're going to cut boneless strip steaks.

- Start by identifying the grain direction and face off one end, cutting perpendicular to the grain. Here, the grain of the loin runs parallel with the muscle. Weigh the whole loin and then determine a portion weight that maximizes your yield. Our loin weighs 4 pounds (i.e., 48 ounces total). A portion weight of 6 ounces will give us an even 8 steaks.

- Measure the length of the muscle. Our loin measures about 10 inches long. Divide that by the number of portions (8) to determine a relative width to cut, in this case around 1¼ inches.

- Measure out the thickness for the first portion, cut, and weigh to determine accuracy. Adjust the width as needed. Cut additional portions, weighing the portion after each cut to ensure accuracy.

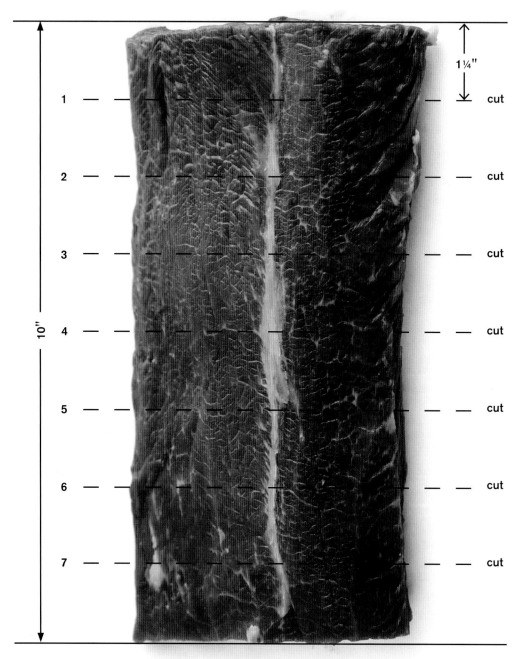

BUTCHERING METHODS

Portioning Based on Thickness

Here we are going to use the strip loin again to demonstrate portioning based on uniform thickness.

- Start by identifying the grain direction and face off one end, cutting perpendicular to the grain. Keep in mind that the grain of the loin runs parallel with the muscle.

- Measure the length of the entire boneless loin and determine an appropriate portion thickness that maximizes the yield. In this example the loin is about 10 inches, so we're going to aim for 1-inch-thick portions, which will give us 10 steaks.

- Measure out the thickness for the first portion and cut.

- Cut additional portions, using the first portion as a visual reference for the proper thickness.

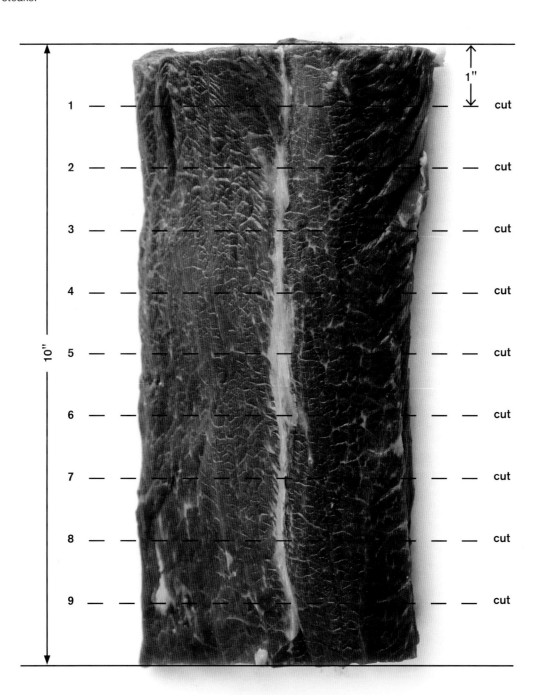

TECHNIQUES 101

Deciding What to Cut

Making your way through an entire carcass cut by cut requires a fair amount of planning to ensure that you are maximizing your yield. It is important to have a clear idea of how each primal will be divided so that portioning and proper storage can happen efficiently, because it may take a couple of days to break down and properly package every cut from the carcass, especially if you are new to the process. No matter how large or small the animal is, the best way to organize your efforts is to use a cut sheet. A cut sheet is a simple concept but is the most valuable tool for minimizing mistakes in processing. It is, in essence, a cutting directive: an obvious process of how to break down the carcass will arise when the required products are clearly laid out.

You can create your own cut sheet by simply going through each primal and writing down a list of what you want to produce from it. After you have read through the butchering chapters in this book, you should know what the pos-

---

Beef Cut Sheet

Name: _____ Customer Number: _____

Address: _____ Carcass Tag #: _____

Farm Name: _____ Dressed Weight: _____ lbs.

Phone # _____ Email: _____
_____

What kind of packaging should we use?   Butcher Wrap _____   Vacuum Sealed _____

Do you have custom labels or stickers that we should use?   Yes _____   No _____

How many steaks would you like in each package? _____   How about for sausages? _____

How much ground meat should we include in each package? _____ lbs.

What would you like the approximate ratio of lean to fat be in your ground beef? _____ / _____

(We recommend an 80/20 mix, lean to fat.)

Do you want your ground meat coursely ground (for stews, ragus, chili, etc) _____ or finely ground (for burgers)? _____

Additional Comments or Special Requests:
_____
_____
_____
_____
_____
_____

VARIETY MEATS
- ☐ Cheeks   ☐ Liver   ☐ Sweetbreads
- ☐ Heart   ☐ Oxtail   ☐ Tongue
- ☐ Kidney   ☐ Spleen   ☐ Tripe

BONES
- ☐ Marrow Bones *Thickness* _____
- ☐ Stock Bones
- ☐ And All the Rest!

SAUSAGES
Hot Dogs: *The classic*   ☐ Qty.: _____ lbs. Linked: _____ or Loose: _____
Beef Salami: *A rustic sausage with fennel, oregano, and anise flavors*   ☐ Qty.: _____ lbs. Linked: _____ or Loose: _____
Spicy Beef Salami: *A spicy variation of our salami with smoked peppers and cayenne*   ☐ Qty.: _____ lbs. Linked: _____ or Loose: _____

sibilities are for the appropriate primals and which ones work for the palates of you or your customers. Create your cut sheet a day or more ahead of when you plan to butcher, to give yourself time to mull over the options and make any changes. Once you cut the meat there is no changing your mind, and what you cut first will determine what else you can cut. One classic example is the short loin: it contains the muscles for strip steaks as well as the tenderloin; in combination they make T-bone and porterhouse steaks. Thus, if you want to cut porterhouse steaks you cannot also have tenderloin and strip steaks, and vice versa. It is therefore important to understand the relationship between cuts when determining your cut sheet. (Fortunately, butchering for yourself has custom advantages: you can cut a few porterhouse steaks off the sirloin end of the loin and then peel off the remainder of the tenderloin and cut strip steaks, giving you some representation of all three categories.)

Beef Cut Sheet

Side 1 (Choose one option under each section)

Side 2 (Choose one option under each section)
☐ Use the same preferences as Side 1

NECK
☐ Boneless   ☐ Crosscut   ☐ Stew Meat

CHUCK
☐ Arm Roasts *Thickness*: ____ (*at least 1.5" recommended*)
☐ 7-Bone/Chuck Roasts *Thickness*: ____ (*at least 1.5" recommended*)
☐ Chuck Steaks (Delmonico, Flat Iron, Shoulder Tender, etc.)
☐ Chuck Eye Roll Roast
☐ Under Blade *Roast?* ____ *Stew Meat?* ____
☐ Ground Chuck   ☐ Stew Meat

BRISKET
☐ Whole *Bone-in?* ____
☐ Point and Flat   ☐ Added to Ground Beef

PLATE
☐ Cross-cut Short Ribs *Thickness*: ____ (*at least 2" recommended*)
☐ Navel *Rolled?* ____ *Added to Ground Beef?* ____
☐ English-style Short Ribs *Length*: ____ (*at least 4" recommended*)
☐ Navel *Rolled?* ____ *Added to Ground Beef?* ____

FLANK
☐ Flap Steaks (Flank, Skirt, Sirloin Flap)
☐ Add to Ground Beef

RIB
☐ Rib Steaks *Thickness*: ____ (*at least 1" recommended*)
☐ Boneless Rib Steaks *Thickness*: ____ (*at least 1" recommended*)
☐ Back Ribs
☐ Standing Rib Roasts *Weight (lbs)* ____
☐ Boneless Rib Roasts *Weight (lbs)* ____

SHORT LOIN
☐ T-Bone & Porterhouse Steaks *Thickness*: ____ (*at least 1" recommended*)
☐ Strip Steaks *Thickness*: ____ (*at least 1" recommended*)
☐ Tenderloin *Roast?* ____ *Filet Mignon?* ____
☐ Bone-in Strip Roasts *Weight (lbs)* ____
☐ Tenderloin *Roast?* ____ *Filet Mignon?* ____
☐ Boneless Strip Roasts *Weight (lbs)* ____
☐ Tenderloin *Roast?* ____ *Filet Mignon?* ____

SIRLOIN
☐ Center-cut Sirloin Steaks *Thickness*: ____ (*at least 2" recommended*)
☐ Sirloin Steaks (Tri-tip, Sirloin Cap)
☐ Sirloin Roasts *Weight (lbs)* ____
☐ Sirloin Steaks (Tri-tip, Sirloin Cap)
☐ Man Steaks *Thickness*: ____ (*at least 2" recommended*)

ROUND
☐ Round Steaks (Santa Fe, San Antonio, Tucson, Western Griller, etc)
☐ London Broil *Thickness*: ____ (*at least 1" recommended*)
☐ Ground Round
☐ Round Roasts (Sirloin Tip, Top Round, Eye of Round)
☐ Ground Round

SHANKS
☐ Whole *Bone-in?* ____ *Boneless?* ____
☐ Cross-cut *Thickness*: ____ (*at least 1" recommended*)
☐ Add to Ground Beef

(Side 2 repeats the same sections and options as Side 1.)

TECHNIQUES 101

Making Knots

If you've ever cooked an oblong roast, a pork chop that won't sit flat on the pan, or even a slice of bacon that curls too much, then you've experienced how the shape of meat can dramatically affect the outcome of cooking it. Evenness of cooking is often a priority, and to achieve that we often need to control the shape of meat. This begins with proper cutting and portioning, but even the best-fabricated cuts will sometimes need reinforcement. Tying with butcher twine is one way to do this.

There are two main knots that I find useful for tying: the butcher's knot and the packer's knot. Both are basic slip knots that rely on a final hitch to secure them. They are virtually interchangeable, so use whichever suits your fancy. In both examples there is a "long" string and a "short" string. The long string will be the one connected to your roll of twine; the short string is the section you are wrapping around the roast. These instructions are written from a right-handed point of view; you'll need to swap hands if you're left-handed.

BUTCHER'S KNOT

Once you get the hang of the butcher's knot you'll understand its main benefit: speed. You can quickly tie and hitch this knot much faster than with the packer's knot. In simple terms, think of this knot as: loop once, knot, loop twice, hitch.

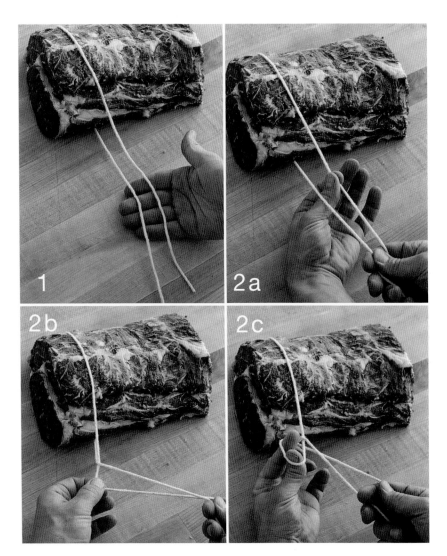

1. WRAP THE ROAST. Run the short end of the string under the roast and then over the top. Hold both the short and long strings in your right hand.

2. FORM THE LOOP. Continue to hold both strings in your right hand during this entire step.

2a. With your left hand, go under the long string and hold the short string between your thumb and forefinger.

2b. Pull the short string under the long string, and twist your left wrist in the process to form a loop.

2c. Your left fingers should be inside, and form, the loop.

BUTCHERING METHODS

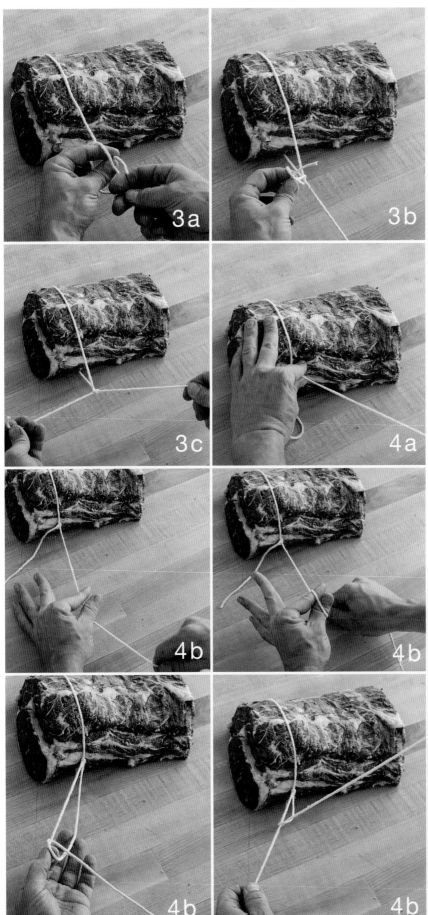

3. FORM THE SLIP KNOT.
Grab the end of the short string in those left fingers and pull it through the loop. This will form your slip knot.

4a. CINCH AND SECURE.
Hold the knot with your left hand and pull on the long string with your right hand to cinch the knot tight against the roast.

4b. SECURE THE KNOT.
Wrap the long string around the thumb and forefinger of your left hand to form another loop. Again, pull the short string through the loop, in this case forming the hitch that will secure the slip knot. Pull the hitch tight and, if you want, make another hitch and double it up.

5. TRIM.
Tighten the knot and trim off the excess string.

TECHNIQUES 101

PACKER'S KNOT

The packer's knot is an attractive knot formed by making a figure eight. Unlike the butcher's knot, which has to be secured with a hitch, the packer's knot has the ability to hold on its own. In simple terms, think of this knot as: under, under, and through the loop.

1. **WRAP THE ROAST.** Run the short end of the string under the roast and then over the top. Hold the short string with your left hand and the long string in your right.

2. **UNDER ONCE.** Pass the short string under the long string from your left hand to your right. There is now an intersection of string, with the short running under the long.

3. **UNDER TWICE.** Run the end of the short up and back under the portion of short string that is laying atop the roast.

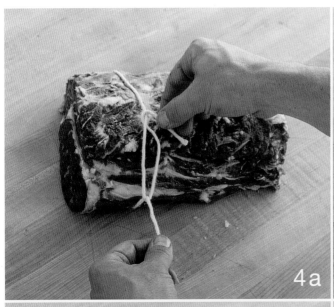

4a. THROUGH THE LOOP AND HITCH.
There is now a loop formed by the short string. Run the end of the short string back over itself and then up through the bottom of the loop.

4b.
Pull on both ends of the short string to tighten, then pull on the long string to cinch it against the roast. Hitch to secure. (See Step 4 of Butcher's Knot, page 71.)

PRE-SLAUGHTER CONDITIONS & GENERAL SLAUGHTER TECHNIQUES

THE RESPECTABLE HARVEST of any animal is less a single event than it is the culmination of many considerations and preparatory procedures taken to ensure the highest-quality outcome. You cannot expect ideal results if the animal has not been tended to properly, equipment is absent or inoperable, additional help is not arranged for when needed, or a cavalier approach trumps the humility of inexperience. It is for these and many more reasons that I impress upon anyone harvesting livestock the obligation of making adequate arrangements, for both the well-being of those creatures destined for death and the perishable products they will provide. You simply can't be too prepared.

The Day Before

THE HARVESTING of any animal will benefit from some basic day-before measures.

Separate Animals

Separate the animals destined for slaughter from the larger group. Unless the slaughter will take place in the field, the holding location for the animals should be near where the slaughter will happen. It's best to give animals 24 hours or more to acclimate to their new surroundings; this will help reduce their stress and anxiety, making handling easier while also helping to preserve the meat quality. The animals' new environment should include easy access to shade and, for animals that need it, shelter. Herd animals like cattle are more comfortable with at least one other animal around, so consider dispatching them in pairs or groups.

PROVIDE ADEQUATE BEDDING, especially if the separated animals are being held in a trailer. Without bedding, animals tend to resist urination, making for a full bladder during evisceration.

DO A QUICK ANTEMORTEM INSPECTION of the animals you'll be harvesting. Should you find anything questionable, call a veterinarian and then decide whether to postpone the slaughter or choose another animal.

Withhold Feed

Plan on withholding feed for a period of time, but ensure that the animals have plenty of access to water. This will help clear their digestive system of fodder and waste, which in turn reduces visceral weight (making handling easier), reduces the potential for contamination (less waste in their entrails), and also improves their dressing percentage (if that matters). Hydrated animals also make for easier skinning, especially if you choose to fist the hide (see page 114). A general guideline for feed withholding time is "the larger the animal, the longer the time," ranging from 12 to 24 hours for veal and beef.

Stunning

STUNNING AN ANIMAL is less about setup than it is about method. In a home setting there is only one way to stun the animal — a blow to the brain — but there are numerous tools available to achieve the same result. Make sure that you have a highly effective approach to stunning, which may include primary and secondary methods, before day-of-slaughter activity commences. Test the

HUMANE HOUSING FOR THE FINAL DAY

All animals deserve to exist in a safe environment without stress and fear, especially during their final hours. Handling and housing the animals with respect is not only dignified and ethical but also essential to producing high-quality meat. (See page 17 for more on the science of stress before slaughter and its effect on meat.)

Livestock animals are timid creatures, susceptible to being frightened by unfamiliar factors. Avoid housing them in areas where they may be subject to startling noises — driveways, construction sites, buildings with loud plumbing — or to moving objects that may startle them, such as fan blades, flags, and wind chimes. Like people, different animals have different needs, so plan on giving them time to adjust to the new surroundings in their own way.

When moving the animals, refrain from inciting fear. Any contact with blunt or sharp objects, such as fenceposts or prods, may cause bruising, even without evidence on the hide. Bruised areas of the meat will need to be cut out, resulting in a loss of product.

equipment far ahead to make sure it works, and provide time for repairs if needed. Accurate aiming is critical; if you miss, you will likely just maim the animal, causing incredible distress and pain in a terrified and unpredictable animal.

Insensibility Is the Priority

When harvesting an animal, our first priority is to reduce the discomfort the animal experiences in its last moments of its life, and for most species this will mean stunning. The main intent of stunning an animal is to render it insensible prior to bleeding. In most cases, the cause of death is exsanguination and the lack of oxygen (via blood) to the brain. To quicken the arrival of death, it's helpful to keep the animal's heart pumping while it's insensible. This will also help promote a more thorough bleed, especially when combined with gravity.

Do not, under any conditions, proceed with the bleeding of an animal when its insensibility is questionable. Extensive research has been done into indicators of insensibility. Dr. Temple Grandin is the leading proponent on the subject; her studies focus mainly on conditions within industrialized abattoirs, but the evidence and guidance contained in her work can be applied to any situation in which animal well-being is a priority during slaughter. The following is a list, prepared by Dr. Grandin, of required conditions to confirm that an animal is insensible and, as you may expect, to confirm the death of an animal prior to skinning or scalding:

- The legs may kick, but the head and neck must be loose and floppy like a rag. A normal spasm may cause some neck flexing, but the neck should relax and the head should flop within about 20 seconds. Check eye reflexes if flexing continues.

- The eyes should be wide open with a blank stare. There must be no eye movements.

- The animal must NEVER blink or have an eye reflex in response to touch.

- The tongue should hang out and be straight and limp. A stiff curled tongue is a sign of possible return to sensibility. If the tongue goes in and out, this may be a sign of partial sensibility.

- In captive bolt stunned animals, insensibility may be questionable if the eyes are rolled back or they are vibrating (nystagmus).

- Shortly after being hung [with your method], the tail should relax and hang down.

- There is no response to a nose pinch or pinprick on the nose. The painful stimulus should be applied only to the nose.

- There is no vocalization (moo, bellow, or squeal) after the captive bolt stunning.

- Rhythmic breathing must be absent. Count as rhythmic breathing if the animal's rib cage moves two or more times.

- When the animal is hung [with your method], its head should hang straight down and the back must be straight. It must NOT have an arched back righting reflex. When a partially sensible animal is hung on the rail, it will attempt to lift up its head. It will be stiff. Momentary flopping of the head is not a righting reflex.

There is no fail-safe method of stunning: at some point you will miss or something else will go awry. Therefore, prepare for an ineffective stun, be it from a misaligned shot, a misfire, or some other aberration. Do not panic if the first attempt does not work. Keep an extra cartridge immediately available — reload, steady the animal, take your aim, and make sure the second attempt is successful. If you are not using a bullet, start from the beginning of the process and quickly deliver a second blow.

A stunned animal will collapse immediately, falling to the ground with all its weight. On hard surfaces like concrete or asphalt, this can result in bruising, especially with larger animals. Therefore, consider where the animal is going to be when it falls and what potential measures you can take to soften the landing, like adding bedding materials to a hard floor or stunning in the grass.

Projectile Stunning

Anyone who has been around livestock, especially cattle, knows that the area in which they remain calm and stationary is limited; this is called the "flight zone." Invading this safe zone will cause the animal to saunter or flee in a safe direction. Commercial slaughterhouses have equipment that restrains the animal (supposedly in a manner that minimizes stress on the animal), enabling the use of close-range stunners like a captive bolt gun. Animals on the farm can be restrained with various types of specifically designed equipment. But retrieving an animal and coercing it into the restraining device is inevitably stressful.

Therefore, at-home stunning for skittish livestock is most often done from afar with a projectile-loaded gun. A basic .22-caliber bullet may work with veal, but something a bit heftier — like a .30-06 or shotgun slug — will be required for older cattle, to reduce the margin of error. It's important to recognize, however, that gun use entails a separate set of conditions. The person conducting the slaughter must be licensed to operate a firearm. For safety reasons, using a gun for slaughter may also limit your options for where to perform the stunning. Even so, a marginally skilled marksman can get close enough to most animals to ensure a positive result and limit any unintended consequences.

If using a projectile method is not an option, then animal restraint followed by a close-range method of stunning is the best approach.

Captive Bolt Stunning

The ideal tool for close-range stunning is a captive bolt pistol, which comes in two varieties: nonpenetrating and penetrating. Both varieties have similar methods of operation. You aim the device at the proper point on the skull and then pull the trigger; a metal bolt, propelled either pneumatically or by a blank cartridge, creates an impact that renders the animal insensible to pain. The effect is instantaneous though not always permanent. Hence, immediately following a proper stun with bleeding is paramount.

NONPENETRATING captive bolt pistols hit the skull with a flat metal disc, causing temporary unconsciousness in the animal.

PENETRATING captive bolt pistols use a narrow metal bolt to penetrate the skull and disrupt brain operations, also causing temporary unconsciousness.

Because the nonpenetrating models are prohibitively expensive, as well as pneumatically powered, the penetrating captive bolt guns are more frequently used in small-scale operations.

Captive bolt pistols are suitable only for animals that can be restrained without excessive duress or anxiety. This will sometimes include calves or mature cattle that have been habituated to close-range interaction. Captive bolt pistols should be operated according to the manufacturer's directions and treated with the same precaution as any other lethal weapon. They must be regularly cleaned and maintained to ensure proper operation and efficacy.

For most larger animals, especially those that are difficult to approach or restrain, the use of a projectile will be the preferred method of stunning.

PRE-SLAUGHTER CONDITIONS & GENERAL SLAUGHTER TECHNIQUES

LOCATE THE STUNNING SPOT

To locate the stunning spot, visually connect the edge of the left horn (or poll) with the right eye and vice versa; the correct spot will be just above the intersection of those lines. Aim too low, and you'll hit the sinus cavity, resulting in a bloody nose and excruciating pain to the animal; aim too high, and you'll puncture the animal's skull but not its cranial cavity. In either case, the result is an animal that is maimed and not unconscious, so take your time.

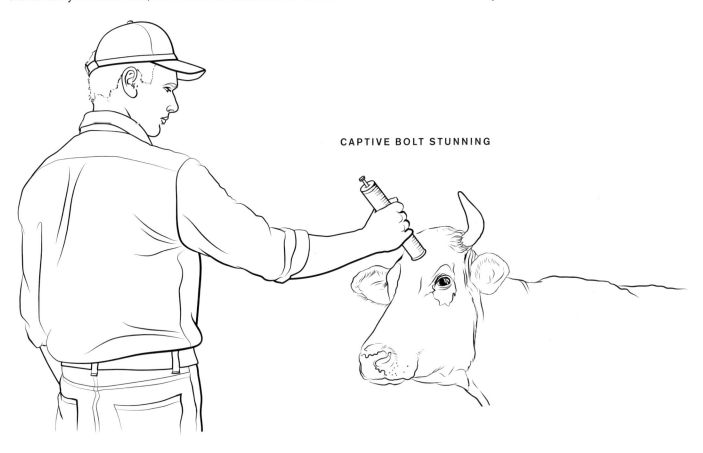

CAPTIVE BOLT STUNNING

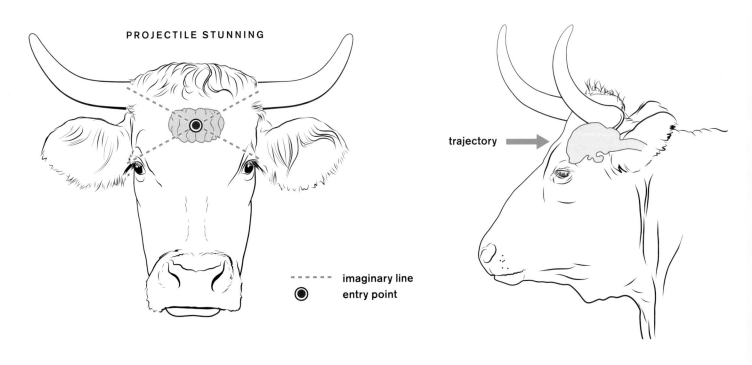

PROJECTILE STUNNING

- - - - imaginary line
● entry point

trajectory

STUNNING 79

Exsanguination

THE VOLUME OF BLOOD that comes from exsanguination is copious. The total percentage of live weight can range from about 4 to 7 percent, depending on the species and age, which for average-size cattle works out to be approximately 6 to 10 gallons of blood (or around 48 to 84 pounds). A successful exsanguination will drain around 60 percent of the blood, leaving you with around 4 to 6 gallons (or 30 to 50 pounds) of liquid that will need to be either saved — for cooking, composting, or creating blood meal for fertilizer — or disposed of. The majority of the remaining 40 percent of carcass blood is contained within the viscera; less than 10 percent will be left within the muscles.

Collecting and Disposing of Blood

Depending on the slaughter location, disposal of blood may be simple. When bleeding an animal in the field, where blood can be left where it lands, be aware of the incline and make sure you're uphill, allowing the blood to flow away from the work area. Handling disposal inside a barn or outbuilding will require more consideration. Floors made from cement, tile, or another cleanable material and fitted with integrated drains are ideal, because they allow blood to be diverted into septic or sewage systems. Regional regulations will apply for this method of disposing of wastewater, a catchall term that covers all the wash water and drainage coming from bathrooms, hand wash stations, kill floors, processing, and other areas of slaughter operations. The reason for this is that wastewater is likely to contain a host of things not normally present in sewage — fats, bits of bone, blood, sanitizing solutions, and microorganisms, to name a few — all of which need to be disposed of in a way that is safe for the environment and the community's water system. Slaughtering for personal consumption, or in low volumes, may fall under exceptions to wastewater disposal regulations, but it behooves you to confirm what requirements you need to meet in your area. If you're unsure about disposing of wastewater in such a way, I recommend collecting it in a container that makes hauling it out to a field for dumping as easy as possible.

Methods of blood collection largely depend on the method of exsanguination and the size of the animal. The larger the animal, the greater risk of injury from death throes, so holding a blood collection bucket next to the throat of a 1,200-pound steer is probably not a great idea. Hanging large animals, those weighing more than 300 pounds or so, over a blood collection vat is the safest approach. This allows you to step out of harm's way while the animal bleeds into a container set below it. Hanging the animal above a wide collection vat also applies to those bled using the transverse method, explained further below, because it involves a large incision across the neck, making it hard to predict the exit point of the blood. The spray area for blood from any hanging animal can be quite broad, especially considering the ferocity of death throes, so the wider the catch container, the better. You can improve your odds of collection by obtaining a wide plastic funnel. You can buy one manufactured for just this purpose, or you can fashion one from a large piece of bendable plastic sheeting.

Transverse Method

The easiest exsanguination method is a clean cut across the neck (i.e., a transverse cut), starting right under the spine, that severs the carotid arteries and jugular veins along with the trachea and esophagus. This method works well and can be used effectively with many species, including cattle. While tests on exsanguination efficiency have shown that blood flow from transverse severing of the vessels is slower than that from the thoracic method, the tranverse method is still popular, simply because it's easy to carry out.

>> See page 100 for a demonstration of the transverse method.

Thoracic Method

The second method of exsanguination severs just the carotid arteries and jugular veins, leaving the trachea and esophagus intact. The blood vessels are cut nearer to the heart than with the transverse method, producing quicker death and a more efficient bleed. The cut is started with a shallow incision from the anterior point of the sternum to the jaw, exposing the carotid arteries and the external jugular vein, which are severed to start the exsanguinations.

When executing this method, it helps to keep the dewlap — the often draping area of skin and muscle between the jaw and sternum — taut. This will help when you make an incision through the hide. If the animal is hanging, the area will become taut naturally; prepare to stand to the outside of either foreleg to perform the incision. If the animal is on the ground, you will want to pull the jaw away from the forelegs to tighten the skin; prepare to stand behind the head to perform the incision.

Testing for Life (and Death)

Death from exsanguination can take anywhere from 90 seconds to a few minutes. You will gain nothing by jumping the gun, so err on the side of caution and wait out the few minutes. With any animal, the eyes should be open; tap the eye to make sure there is no reflex response. There should be no signs of breathing or movement. The tongue should be limp and hanging out of the mouth. If you prick the nose with a needle or knife tip, there should be no response; if you tap the eyeball there should be no blinking. Proceed only when you have confirmed that the animal is no longer living.

Hoisting

ANIMALS ARE HOISTED during slaughter for a variety of reasons, and one is to take advantage of gravity. An animal may bleed more thoroughly when gravity is aiding the heart in expelling blood than when it is lying on the ground. The more complete the bleed, the more resistant the meat is to spoilage. Moreover, when an animal hangs by its hind legs, gravity works on the viscera, pushing them against the diaphragm. This greatly reduces the chances of puncturing entrails when making initial incisions in the groin area. Hanging the carcass also allows you to use both your hands so that you can work quickly and efficiently. Ideal setups for hanging will provide 360-degree access to the carcass; this makes your job easier and quicker while also preventing any contaminated surfaces — walls, fences, poles — from rubbing against an otherwise clean carcass.

Cattle are hung by their feet during bleeding and by their gambrel cords (the animal equivalent of an Achilles tendon) during skinning and evisceration. The gambrel cord is an incredibly strong tendon that can easily support the weight of the carcass as well as any additional downward force applied during slaughtering or

Most farms won't have a mobile slaughter truck like the one we used; a bucket loader or simple rope and pulley setup will work just as well.

butchering. There is a space between the gambrel cord and the calf muscles that allows for the insertion of hooks or rope.

Lifting Options

Bringing the lifting to the animal may be easier than vice versa, which makes a portable method of hanging a huge benefit. A backhoe or a tractor with a loader is an easy and versatile solution. In fact, any mobile machinery that's capable of lifting several hundred pounds can be fashioned to work, as long as the clearance it provides is enough to hang the animal so that it does not touch the ground.

In addition to vehicles, there are dozens of affordable options available for heavy lifting. A block and tackle, a come-along, a winch, and a chain hoist are all good choices. Even a single rope and pulley will work. (Attaching the rope to a vehicle will even save you the effort of having to do the lifting.) These solutions are all relatively stationary, since they need a stable point to elevate from, like a sturdy rafter or a tall A-frame, so you'll need to coordinate where you're going to hang the animal with where you're going to stun it.

With any of these solutions, consider what will be directly above the hanging carcass. A dirty backhoe, a loader used for silage, or maybe a rafter that has never been wiped off? Make sure these areas are relatively clean to prevent contamination from falling debris while processing.

Securing the Animal

Where and how to lift the animal is not the only part of the process to consider — to prevent slippage, a mistake that can result in contamination of the meat or serious personal injury, you must be able to secure the carcass.

GAMBRELS. Any solution for securing the hanging animal should involve two hooks and something that keeps the legs spread to aid in processing. One common method is a gambrel, which is a solid piece of steel with a hook on each end and a centered eyelet for hanging and balancing the weight. You can find gambrels in two formats: a single solid construction or a construction with independently hanging hooks. The former has one major disadvantage: the center eyelet does not allow for imbalanced hanging. This can heavily impact a common scenario: wanting to remove one side of the carcass and leave the other side hanging. For this reason, I highly recommend using the kind with independent hooks. Gambrels are sized for weight and distance of spread; for beef and veal you will need a hefty gambrel specifically designed to support their weight.

SINGLETREE. A singletree is another option that operates identically to a gambrel and is available in the same two versions. Singletrees are most commonly constructed of wood, which is why the meat industry has replaced them with durable stainless-steel gambrels, but for home use a singletree is fine. And because singletrees have fallen out of use, you may be able to get a good deal on a used model.

PARACORD. The animal's back legs can be secured using a heavy-duty nylon twine or cord, often marketed under the term *paracord*. Make sure it has a weight rating well above the weight of your animal to accommodate any additional force. Legs can be tied independently, which is especially helpful if you plan on splitting the carcass. Know your knots before hanging hundreds of pounds above your head!

GRAB HOOKS. If you plan on using a loader or backhoe to hoist the carcass, there is another easy option: a chain

Make sure your gambrel provides adequate spread and support for the animal you are working with.

with grab hooks on either end and a couple of meat hooks, all of which are easy to obtain and affordable. With the grab hooks secured, you can insert the meat hooks into any one of the chain links to determine how far spread you want the animal's legs to be.

Skinning

THERE MAY BE MORE THAN ONE way to skin a cat, but one is all you need to know to get the job done right. Cattle are always skinned, most frequently using a knife; younger veal can also be skinned by hand with a process called fisting. Regardless of which skinning method you choose, there are some tenets to which you should adhere for the sake of sanitation, efficiency, and condition of both the meat and the pelt.

Avoiding Contamination and Damaged Meat

Keep in mind that until you break skin, the interior of an animal is completely sanitary. The only contaminants on a final carcass are those that we introduce, from outside sources or from puncturing the digestive tract or other viscera. Thus, anytime you use a knife while skinning, remember to cut from the inside out. This means that

Designating a clean hand and a dirty hand maintains a clean carcass and speeds up the skinning process.

HANGING DURING BLEEDING

You may choose to hang the animal for bleeding. This is not required, and there are some who assert that hanging before death damages the hind-leg meat by making the muscles conform to an unnatural position and causing a detrimental shift in pH. With nonlethal methods of stunning, do this only if you have a way of hoisting that can bring the animal off the ground in 30 seconds. With anything longer you take the risk of the animal regaining sensibility before the bleeding.

Hanging will involve a separate way of securing the back legs, since the gambrel cords will not yet be exposed and you don't want to make any incisions to the still-living animal. One easy method is a steel chain with an eye on the end allowing for a loop. Have enough length to quickly and easily wrap the loop around one or, better, both shinbones (also known as *cannon bones*). Remove any slack in the chain as you wrap it around the bones; as you lift the animal, the carcass weight will further tighten the chain and secure it. After bleeding, the animal will be lowered for skinning.

making initial cuts through skin is always done with the knife edge facing away from the animal's body, as shown on page 102. This will also save the knife edge: hair and hide are often loaded with dirt and grit, and cutting through the dirty side will dull a knife.

Skinning is as much about removal of hide as it is about producing a clean carcass and avoiding any unnecessary damage from sloppy knife work. Keep your cuts shallow to avoid nicking the muscles, reproductive organs, or abdominal walls. During much of the skinning process, you must keep the hide taut. To avoid damaging muscle tissue, do not pull excessively on areas of the hide that are still attached to the carcass. As you make cuts between the hide and the carcass, the knife edge should be angled slightly toward the hide rather than toward the muscles.

Wet Carcass, Dry Hide

When the hide is removed properly, the interior face should be devoid of membranes, muscle, and fat. It will appear almost dry when compared to a hide that has been improperly removed. This applies to both fisting and knifing — stay above the membrane when skinning. This leaves the most meat, fat, and surface tissue on the carcass, helping prevent excessive evaporation during aging while protecting the underlying muscles from any surface contamination.

Avoid making short cuts while skinning; they cause more uneven surfaces on the hide and carcass, along with carrying a greater risk of puncturing the hide. Keep your knife strokes long and smooth. This will be quicker, cleaner, and produce a better appearance.

Allow gravity to help keep the carcass free from hide grime: the pelt should be removed in such a manner that it naturally drapes away from the carcass. Hanging carcasses are skinned from the top down so that the exterior of the hide can only touch parts of the carcass that have yet to be skinned.

The exterior of an animal's hide is undoubtedly dirty, requiring you to rinse your hands regularly to minimize contamination. You should also designate a "clean hand" and a "dirty hand": hold your knife or fist the hide with your clean hand; handle the exterior of the hide with your dirty hand. Do not confuse the two; wash the clean hand when it gets dirty, and keep the dirty hand away from the exposed carcass.

Evisceration

EVISCERATION IS the removal of the digestive tract and other internal organs, collectively known as viscera. The process of removal accounts for a large portion of the work done during a slaughter. It also poses some of the greatest risk of contamination, since you are working among areas containing digestive and fecal matter.

Avoiding Contamination

Preventing the spread of fecal matter is the first concern during evisceration. Avoid pressure to the abdomen, which can in turn put pressure on the intestines and bladder, causing feces and urine to spill out. With cattle, the anus must be tied shut, preventing spillage when the rectum is pulled down through the cavity.

Again, remember that the abdominal cavity is airtight and uncontaminated until we pierce the lining. Initial cuts should be made with caution, especially because they often occur around the groin, potentially with a full bladder right inside. The walls of the bladder are the thinnest of any viscera you will encounter and, while urine is typically considered sterile until it passes through the urethra, the spilling of it throughout the inside of the carcass is certainly something best avoided.

Evisceration should happen as quickly as possible while maintaining control and a priority of cleanliness. This is especially true with ruminants, whose stomachs are full of partially digested material that continues to be broken down by enzymes even after death. The stomach begins to fill with gases, a by-product of the enzymatic process, and expand the moment death occurs because the gases become trapped. (One step to composting the carcasses of ruminants is to lance the stomach before burial; if this is not done, the gases become trapped and eventually will explode!) Therefore, the longer you wait, the larger the stomach will get and the more compressed the space inside the abdominal cavity will be.

Remember that total cleanliness is impossible; the last step of slaughtering — trimming — only aims to reduce the surface contaminants by removing areas where fecal matter, digestive waste, and other obvious contaminants have landed.

Edible Offal

> The organs you're not interested in eating can be saved and ground into natural dog food.

EXCEPT FOR BONES, most everything within an animal is edible and should be considered before disposal. Cultural tradition is the largest determining factor in what we choose to save for consumption and what we consider inedible or waste. The glands and viscera we choose to eat are commonly referred to as *offal* or variety meats. The organs you're not interested in eating can be saved and ground into natural dog food, a value-added product for those looking to increase revenue per carcass. Any organs intended for future use should be removed from the carcass, cleaned, and frozen immediately. (For more information on freezing organs, refer to page 314.)

The most commonly consumed offal is tongue, heart, liver, and kidneys. In all animals, these organs are routinely saved, even sold individually aside from the carcass or muscle cuts as another revenue source. Beef intestines and bungs, once cleaned, offer edible containers for stuffing minced meat and other goodies into, without which we would never have cased sausage. Chambers of the beef rumen, culinarily known as tripe, are edible after a vigorous cleaning. Even bladders are used for stuffing. Spleens can be grilled, braised, or even simply pan seared.

Oddly enough, even cheeks and oxtail are considered offal these days, despite their apparent external origin; they are both delicious. Brains can be extracted from skulls and cooked, though veal brains are the more commonly consumed, because of the risk of BSE infection from older cattle. Younger animals still have their thymus gland, which is largest at adolescence and slowly atrophies as the animal ages; these glands are better known as sweetbreads and are considered a delicacy. In short, I encourage you to find a way to waste as little as possible.

Brains (Veal)

Brains should only be extracted from veal animals, due to the concern of prions in older animals. The easiest way to extract a brain without damaging it is to partially split the skull in half using an axe or heavy cleaver, allowing you to pull apart both sides and safely remove the undamaged brain. It's also helpful to have a hammer (with the axe) or a wooden mallet (with the cleaver) to improve the cutting accuracy. This can be easily done anywhere as long as you have the tools, so it's perfect for on-farm extraction.

The brain is exposed after the sides are split. Use a knife to carefully cut around the brain, helping to coax it out of its protective cavity. Follow with a good rinse and the following cleaning instructions.

Brains should be handled carefully, since their delicate structure is easily damaged. Peel away the exterior membrane if it was removed with the intact brain. Soak the brains for a day or more in lightly salted water, which helps pull out blood and other impurities. Change the water frequently, every few hours or so, after which the brains will be ready for cooking.

Oxtail, chock full of collagen that hydrolyzes to gelatin, will provide depth to any stew or stock.

Cheeks

Cheeks, also known as the *masseter* muscle, are found on all animals and responsible for mastication. Cheeks from ruminants like cattle will need to have salivary (parotid) glands removed. They exist in clusters and are clearly identified by their beige-to-brown coloration and spongy texture. Use your knife to slice them away while maintaining a relatively flat surface for the cheek or jowl as a whole, avoiding the unnecessary removal of any meat.

Heart

Sever any remaining tubes attached to the heart as well as the pericardium, the sac that surrounds the heart. Cut open the heart by slicing partially through the top, exposing the chambers and allowing you to expel any blood clots by squeezing. Rinse thoroughly, inside and out.

The hearts of large animals like cattle require some additional cleaning prior to cooking. Cut the heart into manageably flat pieces, often corresponding to chambers. Trim all the exterior fat from the pieces, leaving a clean surface while trying to remove as little of the heart as possible. Flip the pieces over and trim away the tough interior membrane that lines the chambers. Cardiac muscle that is ready for cooking should be dark with a dull finish.

Kidneys

Kidneys require only a thorough cleansing prior to storage. Ensure that the kidneys have been "popped" — that the outer translucent membrane has been removed. (See page 122.) Also, slice them lengthwise along their midline and remove the fat deposits within.

Liver

The liver needs minimal cleaning in order to be palatable; mainly, removal of the gallbladder while working to avoid spilling any bile on the organ. Remove the gallbladder by pinching the duct where it connects to the liver and slowly tear the duct away from the liver in the direction of the gallbladder. Stubborn duct connections can be severed with the tip of a knife. Look for signs of parasites or infection such as dead or alive flukes and white lesions; if you find any, then you must dispose of the liver.

Spleen

After you trim away any membranes or connective tissue, thoroughly wash the spleen and store it frozen until use.

Sweetbreads (Veal)

The sweetbreads require the same preliminary steps prior to cooking whether they be from the thymus or the pancreas. (True sweetbreads, from the thymus gland, can only be gathered from veal animals. The gland atrophies as the animal ages.) Rinse the sweetbreads in a bowl of cold water, gently cleaning the surface of any collected residues or blood. Soak them for up to two days in lightly salted water or milk, which cleanses the glands by drawing out stagnant blood and impurities, changing the liquid every few hours. After a good soaking, briefly poach the sweetbreads and then shock them in ice-cold water. Once they are cooled, peel away the exterior membrane, including any veins or gristle, leaving them recipe-ready.

Tongue

Thoroughly rinse the tongue, scrubbing the papillae-covered surface with your hand or a brush. Tongues need to be peeled before they can be consumed. To do so, remove the tough, tastebud-lined surface by poaching the tongue for 2 to 3 hours, depending on the size of the animal. After poaching, remove the tongue and let it cool until it can be handled. The thick outer layer should separate quite easily; if not, poach longer and repeat.

Considering the somewhat tedious task of cleaning hearts, the beef heart returns in spades for your efforts, providing a considerable amount of fantastic-tasting meat that is easy to prepare. (Since it never stops working, it may very well be the most flavorful part of the animal.)

Carcass Cooling

PROMPT COOLING OF ANY CARCASS is imperative for limiting microbial growth, thereby preventing spoilage and a loss of product. You'll need a meat thermometer to test the internal temperature before further processing. The target temperature for the carcass is 36°F to 40°F, and the ideal time it takes to get there is dependent on the species and size; large carcasses can take up to 36 hours. Faster is not better in this case, because quick cooling carries the risk of cold shortening. (For more on cold shortening, see page 17.)

All carcasses benefit from a period of aging. The specific length of time is determined by size and personal preference. Following the cooldown period, proceed with aging based on the methods described on page 18.

A Quick Rinse

Before heading into the final cooling process, it's beneficial to rinse the carcass. Carcasses headed for cooling in an environment with good air movement — e.g., a blast cooler or cold space with fans — will benefit from a quick rinse with cold water sprayed from a hose. All carcasses, regardless of cooling environment, benefit from a vinegar rinse, which helps fight any pathogenic growth on the surface. This can simply be a 1-to-1 mixture of vinegar and water sprayed on the carcass with a quality spray bottle. Go light on carcasses that will cool in environments with little to no air movement — e.g., large chest coolers — as excess moisture can be problematic for the aging.

The most difficult stage to accommodate in large-animal processing is the cooling down and aging process, simply because of the size of the animal. The animal needs to come down to below 40°F in less than 36 hours. That requires either a cold day and night (nature-assisted cooling) or a large walk-in cooler (mechanical-assisted cooling).

NATURE-ASSISTED COOLING

Nature-assisted cooling can happen when the temperatures for an outdoor enclosure range from 26°F to 40°F. Anything outside of that range will cause either spoilage from heat or freezing from cold, both of which should be avoided. If nighttime temperatures are expected to fall slightly below 26°F, you can wrap the carcass with cheesecloth or muslin to prevent any considerable freezing. Obviously, this temperature range is hugely dependent on your local climate, but for areas that have cold winters this range usually will occur in the late autumn or late winter/early spring. Choose a clear day without precipitation or high winds to help reduce excess moisture and airborne contaminants.

Nature-assisted cooling also presents challenges for aging. How consistent the temperature will be over the weeks following slaughter is uncertain. Keeping the carcass enclosed to prevent attracting predators or scavengers may be difficult. With this method there is no real control of humidity, which is usually much lower than the ideal range. Do your best to control the environment for aging; if, however, conditions become detrimental to the meat yield, it is time to butcher.

MACHINE-ASSISTED COOLING

Machine-assisted cooling is an ideal option for those wanting to slaughter year-round or in areas where temperatures never get as low as needed. Walk-in coolers should maintain a temperature of around 30°F for ideal cooling and 34°F to 38°F for aging; a relative humidity of 70 to 80 percent is ideal in both circumstances. Make sure the distance between the ground and the hooks or other hanging implements provides enough clearance to accommodate the whole, halved, or quartered carcass. You don't want any part of the carcass touching the ground or lying on its side while it's cooling or aging. Allowing for ample air circulation around the whole carcass will decrease cooling times, reduce bacterial growth, and prevent overall spoilage.

No matter which method of cooling you choose, ensure that it's progressing at a satisfactory rate by taking a temperature reading every 12 hours or so. Test by inserting a thermometer into the thickest part of the thigh meat, near the bone.

BONE IT WHILE IT'S HOT

It is highly recommended that all carcasses go through the proper cool-down process. There is no shortcut for food safety, and this is one of the most important steps. Yet the carcasses of larger animals take up a considerable amount of cold space, and this can be hard to come by. Hence, quickly breaking down a carcass into primals prior to cooling may be the only viable option.

Breaking down a warm carcass is known as *hot boning*, a process generally not recommended for home butchery, for a couple of reasons. The temperature of a warm carcass falls right in the middle of the "danger zone." (Body temperatures are ideal for pathogenic propagation.) Therefore, the breakdown process into constituent parts must be swift and efficient, preventing any excessive delays before the meat makes it into the cooler; the butcher's handiwork must be deft, and many home butchers are not accustomed to working at such speeds. Furthermore, the ambient temperatures within home processing environments are often higher than ideal (40°F), doubling the risk of delays prior to cooling. Nonetheless, some home butchers may have cold storage restrictions that limit their options; fitting a whole beef quarter in a stand-up refrigerator is impossible, for example. In these cases the carcass must be promptly processed into primals or subprimals, and then packaged for cooling and aging. The process of primal and subprimal breakdowns is included in each butchering section and should be followed for hot boning as well. Keep in mind that primals and subprimals can be further broken down into portions and cuts after they have reached a safe temperature, 32°F to 36°F.

Here are some recommendations if hot boning is your required course of action:

- Sanitize your tools and processing station ahead of time. Sanitation, while always important, is critical when working with a warm carcass.

- Set up your processing environment before you begin the slaughter. This includes having packaging ready and ensuring the final cooling environment is at its target temperature.

- Bring the ambient temperature of your processing environment as close to 40°F as possible.

- Bring in some help and set up workstations. For example, you perform the initial butchering into primals; someone else butchers primals into subprimals; the next person packages them; and the final station weighs, labels, and deposits in the cooler. An assembly-line approach will dramatically speed up the process and get items into the cooler quickly.

- Immediately begin hot boning after the slaughter is complete.

- Use less force than usual when separating muscles and exposing seams. Warm, un-aged muscle is much more delicate than its cooled counterpart. Tearing can easily happen, irreparably damaging muscles.

- Do not immediately freeze hot boned meat, which runs the risk of cold shortening (see page 17); it still needs to age, after which it can be frozen. (Aging times are found on page 18.)

Cleanup and Disposal

PROPER POST-SLAUGHTER CLEANUP is an important step; you should leave the work area free of blood, entrails, or other material that will attract predators or produce foul odors.

Disposing of Inedibles

First deal with the inedibles: entrails, blood, and feathers. There are myriad options for disposing of slaughter detritus, but the most common are compost, trash, or a rendering company. Do not dispose of animal waste by dumping it and leaving it for scavengers. This is a high-risk solution with regard to disease transmission on your farm and poses risks to the safety of the surrounding water, which in turn endangers not only your livestock but also the surrounding wildlife and any pets in the area. Moreover, attracting scavengers can result in higher predation on your more vulnerable animals.

COMPOSTING

Composting is by far the best option. It is the cheapest and least wasteful, and it results in a nutrient-rich compost. It will require having wood chips, sawdust, and other organic material on hand, but with a bit of forethought you can easily acquire the necessary materials. If you're processing any species on a regular or even semi-regular basis, it will definitely help to maintain a compost pile; such piles are relatively self-perpetuating with the right maintenance.

TRASH

Disposing of inedibles within a municipal trash system is another option. First, double-check with your town or garbage company to ensure that what you intend to dispose of is allowed. If so, then choose thick plastic bags, often called contractor bags, to help prevent punctures and leaks. Also, make sure the area where the bags are left for pickup is secured from scavengers and other animals looking to bury their nose in a fresh or not-so-fresh kill. Or place remnants in metal garbage bins with heavy weights — like cinder blocks — on top to prevent inquisitive creatures from getting inside.

RENDERING

Most likely, there's a rendering company in your area that aids farms in the disposal of sick animals, placental membranes, butchering residue, and other renderable matter. Prices and requirements for handling the waste differ — many companies do not allow rumen manure, for example — so make sure to contact the company well in advance to allow for any preparations, like procuring a 50-gallon drum to hold the waste.

STORING AND MOVING INEDIBLES

Moving the inedibles of larger animals is going to be more difficult. A loader or backhoe will make moving 200 pounds of guts an easy task. A good-size garden cart or pushcart lined with plastic (to avoid staining and other additional cleanup measures) may also work, depending on the terrain you need to cover. As a last resort, a wheelbarrow can do the work, but it may take a few trips. After the animal has been hoisted, lining the ground beneath it with a durable plastic tarp will help keep the area clean and can also help in collecting and moving the waste.

Work Area and Tools

Your work area should be cleared of all blood and other waste, which may attract vermin, especially if it is in an enclosed area or near where you're hanging the carcass for cooling and aging. After cleaning, spread sawdust to absorb any remaining liquids; the sawdust can then be shoveled to a compost pile or discarded.

All the equipment should be washed thoroughly with hot water and soap, and then sanitized before being stored. Anything not made of stainless steel should be hand-dried to prevent corrosion. Knife steels should be completely dried immediately after cleaning to prevent rust buildup. (The capillary action of paper towels works best for drying grooved steels.) Other cutlery can be left to dry naturally.

SLAUGHTERING ESSENTIALS

5-inch boning knife

6-inch boning knife

Skinning knife

Honing rod (ceramic)

Captive bolt

Bone saw

Tools and Clothes for Slaughtering

THE HAND TOOLS REQUIRED for slaughtering are quite minimal, especially when compared to those required for additional processing.

Knives and Other Tools for Slaughtering

Knives should be recently sharpened and should come with handles constructed from a material that makes sanitization an easy task.

A KNIFE FOR ALL-AROUND TASKS should have a 5- to 6-inch semi-stiff curved blade. I prefer one of these for all-around tasks when harvesting species other than poultry.

A SKINNING KNIFE has a wide and severely curved blade. It will help in the skinning of larger animals, but it's not required to do the job well.

A HONING STEEL is needed wherever there are knives. Your knives will dull quickly, especially when you're trying to cut through ligaments and cartilage or removing the hide.

A SAW, ideally intended for cutting through bone, is needed to get through the sternum. Electric meat saws are excellent but costly. Handsaws are another option that will get the job done. Look for a saw with a blade length of 23 to 25 inches.

TRIM HOOKS can aid the process of cutting away small areas of contamination from a carcass before it heads into the cooler.

STURDY TWINE should be kept on hand. Cut an 18-inch length for tying off the bung.

Clothes

When referring to the accumulated muck from slaughtering and butchering as compared to the typical daily grime, a mentor of mine once said, "There are different kinds of dirty." Expect any clothes you wear to get heavily soiled with blood, feces, fats, and everything else around. Rubber aprons, similar to those worn by restaurant dishwashers, are easy to clean and can be purchased online. Cloth aprons work too, though staining is permanent. Sleeves should be rolled up to the elbows to avoid smearing contaminants anywhere your arm goes.

Footwear should be nonslip if you're working on tile or other smooth surfaces.

Nitrile or latex gloves are helpful to have for a few reasons: you may not always want to work with bare hands; the tackiness will help you grip slippery surfaces; nicks and wounds happen, and when they do you should cover your hands to prevent contamination.

Required Space

IT TAKES SURPRISINGLY LITTLE SPACE to process even the largest animals. The equipment you'll use may be the main factor in figuring out what your space needs will be. Animals that can be stunned, skinned, and eviscerated all in the same space require very little room. However, if you're going to use a tractor to lift an animal, the tractor — not the process — will define the amount of space you need. Determine what equipment you will use, especially for hoisting, and that will determine how much space you need. When working in an outdoor space, pick a location where the ground is flat and clean, has good drainage, and isn't prone to excessive dust.

Access to potable water is a requirement. Running water is helpful for spraying carcasses prior to cooling and cleaning offal. Keeping yourself and the carcass clean during slaughtering will require repeated hand washing, so plan ahead on having a hands-free solution: options include two buckets, one with soapy water and one with clean rinse water; a foot-operated faucet; or even a secured hose with a ball valve. Try to get plastics — buckets, cutting boards, cooling containers, and hoses — that are food-grade.

6

SLAUGHTERING CATTLE

BEEF IS BY FAR the most popular red meat produced and consumed in the United States, with the average American consuming more than 60 pounds of it each year. Since the average yield from a whole carcass of beef is more than 400 pounds of boneless meat, a single animal has the potential to provide an entire year's worth of beef for a family of seven. Furthermore, frozen beef stays palatable for a longer period of time than almost any other kind of meat, allowing you to extend your harvest well beyond one year. This simple equation may be reason enough for you to learn how to slaughter your own cattle on-site. Fortunately, the process of slaughtering cattle is uncomplicated, and the slaughter of a single animal can be easily completed in one morning.

Setting Up for Slaughter

HARVESTING CATTLE ON THE FARM is the final punctuation of a process that will have started many months (or even years) before the day of slaughter, so it behooves you to ensure that all the necessary arrangements for slaughter will have been made well in advance. Not only does the animal need to be in prime condition, but you also need to be prepared for both the process of slaughter and the result: a large carcass that will require lengthy storage, followed by considerable amounts of fabrication (cutting into usable portions), packaging, and freezing.

Fortunately, slaughtering cattle is actually much less complicated than you might envision; the slaughter process for cattle involves fewer stages than for pigs or even chickens. So, while the size of a cow needs to be considered, do not be intimidated by it. Nonetheless, do not try to go it alone. At minimum, include one other (strong) person to aid in the process, and if this is your first time, try to include two or three people. Properly prepared for, the slaughter process will go smoothly and can take less than 45 minutes for a single animal. It need not be an all-day affair.

Information on setups and other considerations for beef are covered in chapter 5.

The Right Age

Determining when to slaughter cattle and how to grade them is, quite literally, a science. The influencing factors are vast: age, breed, feed, season, genetics, conformation (or body shape), and more. The conventional meat industry typically sends young cattle, called feeders, to a feedlot to be bulked up. Feeders are around 12 months of age when they arrive at a feedlot. They spend the next 150 to 250 days or so feeding on a specially composed diet to increase carcass yield and impart specific characteristics, most frequently marbling, that producers look for in the final product. The United States Department of Agriculture (USDA) has developed its own system of grading, based primarily on marbling, which is the interspersal of fat within the muscle.

For a small operation, deciding when to slaughter cattle will, most likely, be far less complicated. Most commonly, beef slaughters are performed during the colder months, which allows for extended periods of outdoor aging without the need of a large, walk-in cooler. However, temperature and season alone cannot denote when an animal is ready to be slaughtered: age and weight will be important factors to determine when your animal is

---

EQUIPMENT OVERVIEW

It's surprising how little equipment you actually need to properly slaughter cattle. You can do it bare bones: with a knife, a handsaw, and a hanging implement. However, there are some tools that help smooth the process and speed things up. Here is a list of the slaughtering activities and their associated equipment options, listed in order from minimal to ideal:

STUNNING: bullet gun, captive bolt gun

HANGING: block and tackle, come-along, tractor/backhoe, electric winch

BLEEDING (EXSANGUINATION): 5-, 6-, or 7-inch knife

SKINNING: 5- or 6-inch knife, skinning knife

ESOPHAGUS SEPARATION: weasand rod

EVISCERATION: 5- or 6-inch knife

CHILLING AND AGING: well-ventilated outbuilding (ambient temperature dependent), blast chiller walk-in cooler

ready. The weight and conformation of the animal will be largely breed dependent. As a broad generalization, British cattle breeds (e.g., Angus, Hereford, and Shorthorn) will finish at a lower weight and will have more pervasive marbling than cattle of continental breeds (e.g., Charolais, Limousin, and Semmental), which tend to be leaner animals with larger frames and higher final weights. Older animals will tend to provide a richer, deeper flavor, because their muscles have been active for longer periods of time as compared to younger cattle (see page 9).

Preparation

Preparation for the slaughter should begin the day before the undertaking. Most important, the animal should be separated from other animals and food withheld for around 24 hours, giving the animal some time to settle into its new surroundings. Even before that, make sure you know exactly how you will proceed through each step, especially the stunning and exsanguination, which happen very quickly. Your first time through the slaughtering process can be disorienting, so thorough preparation is essential. Ensure that all equipment has been tested and is working, and make time in your schedule to accommodate for repairs, if needed. Read through chapter 5, as it covers the foundations required for each stage of slaughter, without which the process is bound to be fraught with error.

The day before slaughtering, separate the animal from the herd and withhold feed.

RAISING VEAL HUMANELY

Today's commercial veal market is almost exclusively a by-product of the dairy industry: male calves provide little to no use for dairy operations, which rely on females for milk production, and so they are more often sold to veal operations that raise the calves for early slaughter. The typical characteristics of veal — lean, tender, milky-colored meat — are produced through confinement and a low-iron diet, which causes anemia and provides the unnaturally light hue to the meat. Understandably, there has been an ethical backlash against the disrespectful conditions of commercial veal production.

Farmers, often working with smaller operations, have stepped in and have begun providing a respectable alternative to local markets: pastured, or rose, veal. These animals are not confined but are raised with their mothers before being put on pasture for the final few months of their lives. The meat is a deeper rose color — hence the name "rose" veal — due to its healthy diet. The meat also develops more flavor because the animal is free to move about on pasture and experience natural behaviors. Thus, for the first time in modern agriculture, veal is being offered in local markets to conscientious consumers.

A veal calf is a lean animal because it has not been given time enough to bulk up in weight and fat. Nationally, the veal calf is classed in terms of size and weight, which ranges from weeks to months in age and up to around 400 pounds. Home processors will typically process veal calves under three weeks of age as bob veal, or after a few months as pastured, or rose, veal. The color of the meat will be based on the diet of the animal, be it milk, formula, specially designed veal feeds, grains, or grass.

SKELETON AND VISCERA OF CATTLE

Cattle are the largest species of livestock commonly found on American farms: a mature cow can weigh well over 1,200 pounds, while an average pastured veal calf may weigh as much as 400 pounds. While cattle are far larger than most other four-legged livestock species, their skeletal structure still shares many similarities with them.

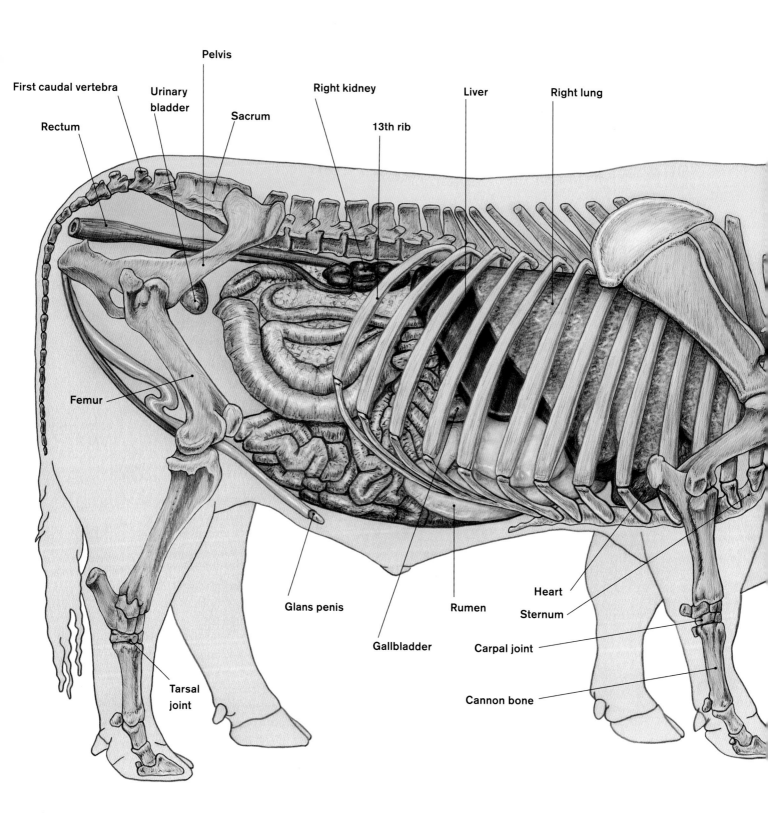

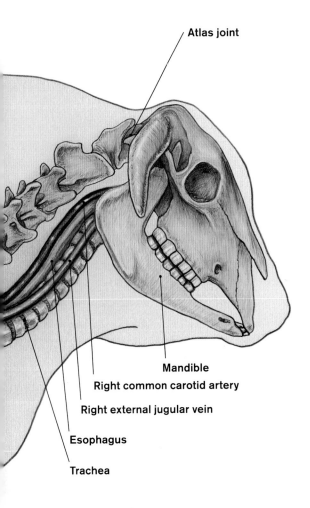

- Atlas joint
- Mandible
- Right common carotid artery
- Right external jugular vein
- Esophagus
- Trachea

THE MAIN BLOOD VESSELS – *carotid arteries* and *jugular veins* – cross as they leave the heart. This intersection occurs right around the front (anterior) edge of the sternum. Severing the vessels at this junction is the main priority when using the thoracic method of bleeding.

THE ESOPHAGUS AND TRACHEA leave the skull and run side-by-side until they enter the thoracic chamber. The blood vessels that need to be severed run above on both sides of these tubes. Severing just the blood vessels at the base of the skull is far too difficult; thus, the esophagus and trachea are severed along with the blood vessels when using the transverse method of bleeding.

THE ATLAS JOINT connects the base of the skull to the *first cervical vertebrae (C1)*. It is this connection that must be disjointed in order to remove the head. For horned animals do this early on before the skinning process reaches the skull. Don't forget about the tongue and cheeks, which should be removed before discarding the skull.

THE CARPAL JOINT is the lowest joint in the forelegs. You remove the front hooves here. The *tarsal joint* is the lowest joint in the hind legs. There you will remove the hind hooves and, as it is far more difficult to disjoint than the carpal joint, you may find yourself using a saw.

THE STERNUM will need to be split to aid in the evisceration of the thoracic chamber. The posterior (back) tip is made up of cartilage – as you saw through it, be wary of going too deep and nicking the *rumen* or other abdominal viscera.

THE PELVIS, or aitchbone, once split, is the bone you'll need to tackle when separating the bung, or rectum. Splitting the pubic symphysis, the cartilaginous midline on the bottom (ventral) edge of the pelvis, is one method for separating the bung.

THE SPACE BETWEEN the *twelfth and thirteenth rib* is where to cut when quartering a carcass for hanging. Make sure to cleanly knife through the loin muscle to avoid tearing as you saw through the spine.

CHEEKS, TONGUE, HEART, SPLEEN, LIVER, KIDNEYS, and **VEAL BRAINS** are the likely candidates for offal saved after slaughter. The first three chambers of the *rumen* can also be cleaned for tripe. The more reactive organs, like the brain, liver, and spleen, are best when frozen immediately after slaughter, though all offal benefits from this approach.

THE SLAUGHTER PROCESS

Everything about cattle is big — the land they require; the efforts and cost to raise them; the equipment necessary for slaughter; and the volume of meat they return. Certainly, their massive size is going to be the greatest challenge to accommodate during slaughtering, but fortunately the process is straightforward and requires little relocation of the carcass until the final step, which is cooling. You must make a couple of key choices when slaughtering cattle. The main choice involves whether to skin the animal while it's on the ground or to hoist the animal and skin it while it's in the air. You must also decide whether or not to separate the esophagus from the trachea by using a weasand rod. Regardless of which of the previous methods is used, the slaughtering tasks will most likely occur in the following order. Note the steps that occur while the carcass is dirty, on the left, and those that will require more consideration of cleanliness after the carcass hide has been largely removed.

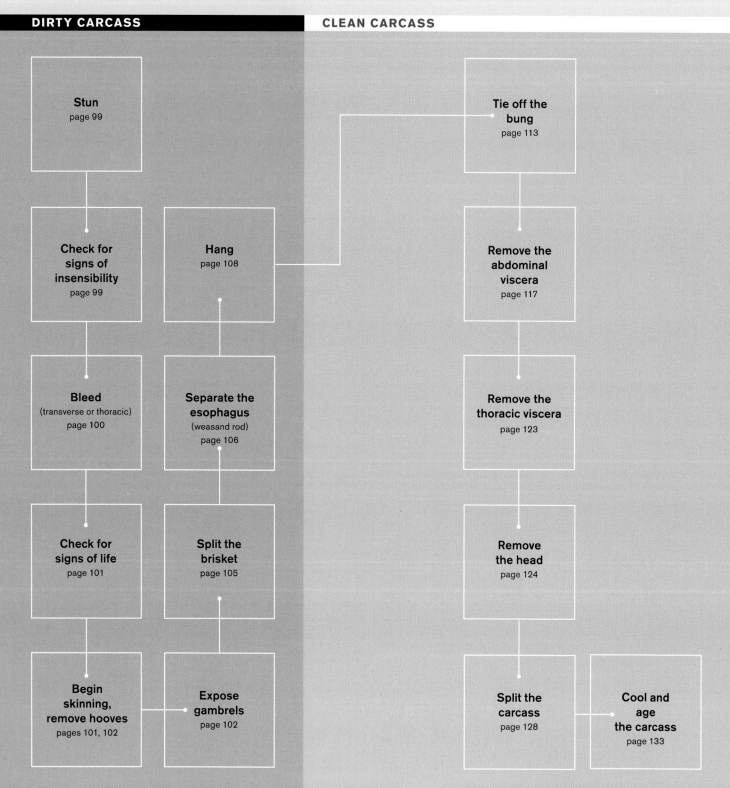

Stun the Animal

Stun your animal only when you know it's experiencing a very low level of stress. Regardless of whether you choose to stun it with a projectile gun or a captive bolt gun, the area to aim for is the same (see Locate the Stunning Spot, page 79). Stunning the animal in this spot will produce immediate unconsciousness, thus desensitizing the animal during the process of exsanguination (which is the actual cause of death).

Stunning calves should follow the same guidelines as full-grown cattle. The area to aim for is the same, but the skull of a calf is far less resilient, due to its younger age, so achieving a successful stun requires less force of impact. Nonetheless, it's best to err on the side of caution: when preparing to stun a calf, choose a method of stunning that would be appropriate for full-grown cattle.

Once you and the animal are ready, proceed with your planned method of stunning (in this case, a rifle). With an effective stun, the animal will immediately drop to the ground. Without delay, check for signs of insensibility (see page 77).

Bleed the Animal

Once the animal is down, test for insensibility (see page 77). Then proceed with bleeding according to the directions for the method you've chosen (see page 80). Here we are using the transverse method.

Situate the Carcass for Skinning

Proceed to skinning only when you have confirmed that the animal is no longer living. Follow the recommendations outlined on page 83 for knife work when skinning.

Grab hold of the front and/or back legs, and then use the legs to roll the animal onto its back, legs pointing straight up. Situate some sturdy objects on either side of the carcass, nestled close to the spine, to provide stability and prevent too much lateral movement. These objects should sit low to the ground and should allow the hide to drape away from the body as it is removed. You can use any relatively heavy objects to keep the animal propped up: tires or cinder blocks both work well. These objects do not need to be overly clean or sanitized; they will not contaminate the carcass because they will only be in contact with the already-soiled hide.

Remove the Hooves

The hooves are by far the dirtiest part of the animal, so remove those first. To speed up the process, one person can work on the front while another person works on the back. Each front leg can be held steady by pushing down on the hoof with your dirty hand while the knife hand makes the initial incisions. (Remember to clean your hands frequently during these initial stages because the soiled hide is chock-full of contaminants.)

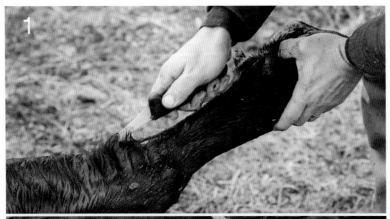

1. **SKIN THE FRONT LEGS.**
Slide your knife just below the surface of the skin, starting at the hoof and ending just past the carpal joint. Knife the skin off both sides until you expose the carpal joints enough to knife through them cleanly.

2. **SEVER THE LIGAMENTS.**
Using the tip of your knife, sever the connecting ligaments while applying downward pressure on the hoof. (It may prove challenging the first time, but with practice it will become a simple step.) Once the joint pops open, knife through the underside connections. Using a handsaw is also an option if knifing proves difficult.

3. **REMOVE THE REAR HOOVES.**
Approach the rear hooves in the same manner as the front, peeling away the hide to expose the tarsal joint. As you skin the rear legs, make sure to cut on the inside of the leg and peel the skin back, allowing it to drape away from the leg. Knifing through the tarsal joint is much more difficult than knifing through the carpal, due to the shape of the fused tarsal bones and the density of overlapping ligaments. (If you run into trouble, pull out the handsaw and lop off the hoof.) Above all else, be careful not to sever the gambrel (Achilles) tendon or its attachments, as this is where you will insert your hanging implement and no other tendon will suffice. Once the legs are skinned, find the space between the gambrel cord and tibia; insert your knife and cut parallel to the gambrel cord, making just enough space for hanging implements.

Begin to Remove the Hide

Now that the hooves are removed, you can move on to the main hide removal. Connect the back leg incisions across the groin. If you are working on a male, be careful not to sever the pizzle (penis) which can leak urine. Slide your knife underneath the hide and cut along the entire middle of the carcass. (Again, if the animal is a male, avoid the pizzle by keeping to one side or the other. If you are working on a female animal, start your incision behind the udder or go around either side of it.) To avoid puncturing the abdomen, pull the hide away from the body as you cut. Once the midline of the hide is removed you can begin "siding" – the process of removing the side hide – while the carcass is on the ground.

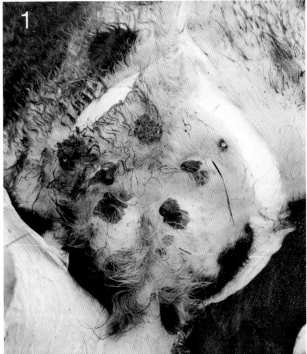

Stop when you've passed the point where the abdomen begins to slope back toward the spine. (You can go farther if you do not need to kneel in order to do so.) Remember to use long, smooth strokes for each cut. One hand keeps the hide taut, pulling up and outward, and the other (clean) hand cuts, avoiding cutting holes in the hide while keeping the protective outer layers of fascia and meat on the carcass. Clean your hands frequently during this process to avoid contamination.

1. **CUT AROUND THE UDDER** when making your centerline cut to start siding.

2. **TEAMWORK HELPS** make siding a quick process.

3. **KEEP THE HIDE TAUT** and run the knife against it with the blade angled slightly outward.

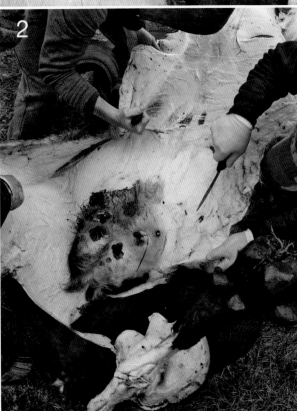

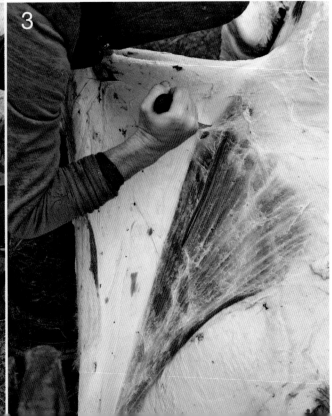

Separate the Genitals

Genitals can be separated anytime after the initial centerline of the hide is cut. For bulls, the testicles should be removed. Hanging testicles can be easily removed by cutting through the point where they join the abdomen. Proceed carefully to avoid puncturing the abdomen or severing the pizzle (which can leak urine).

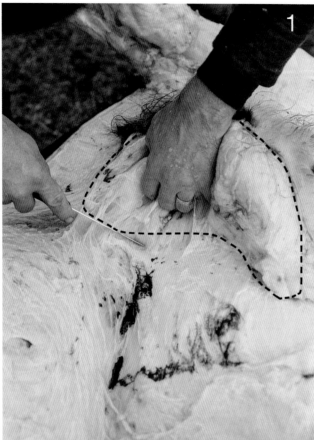

Removing the mammary glands and teats from a female is a bit more difficult than removing testicles from a bull. Find the upper edge of the mammary gland – you can feel the softness of the gland underneath the hide, as compared to the firm feeling where the hide lies atop the abdominal wall. The gland is surrounded by fatty tissue and connected by a tendon (called the symphyseal tendon). This tendon is less ropelike than the tendons found on skeletal muscles and is more like a sheet of connective tissue. Removing the udder is distinctly different from removing hide from muscle. When you hit the symphyseal tendon, continue cutting but proceed slowly, ensuring that you do not cut too deeply and open the abdominal wall. Using the abdominal wall as a guide, follow the surface around both sides of the mammary gland and finish separating it near the groin.

1. **CUT THE FATTY MASS** of the udder away by following its seam with the contour of the abdominal wall.

2. **RELEASE BOTH SIDES** of the udder and then begin to sever the symphyseal tendon, which is tough and runs medially.

3. **SEVER THE TENDON** at the groin last.

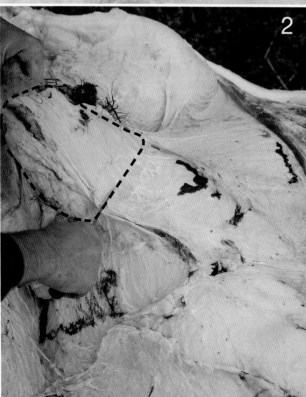

Split the Brisket

Once you have removed the hide from the brisket, you can start splitting the brisket and sternum. Doing this while the animal is on the ground, rather than while it is hanging, is advantageous for a couple of reasons. First, it's easier to saw through the sternum horizontally rather than vertically. Second, the abdominal viscera, which lie right behind the posterior tip of the sternum, lie against the spine rather than the sternum and diaphragm, as they do when the animal is hanging. This placement of the abdominal viscera against the spine helps prevent accidental punctures of the rumen with the saw.

1. CUT THROUGH THE FAT AND MUSCLE.

Find the midline of the sternum and cut through the fat and moat to expose the bone. (Remember, it's always better to cut through meat with a knife than a saw.) Start at the posterior end. As you cut, keep in mind that the posterior tip of the sternum is cartilage and could be split, puncturing the abdomen, if too much pressure is applied. Make one pass with a shallow cut, and then make another pass cutting down to bone.

2. CAREFULLY SAW THROUGH THE STERNUM.

Stand on the side of the carcass or directly in front of the sternum when splitting the sternum. Using your longest bone saw, start sawing where the sternum slopes toward the neck, keeping the angle of your saw at about 45 degrees, or to an angle that is perpendicular to the slope of the brisket. Then saw along the flat of the sternum. In order to avoid puncturing the rumen, keep your saw at a low angle, maybe 20 degrees, being careful not to cut too far back.

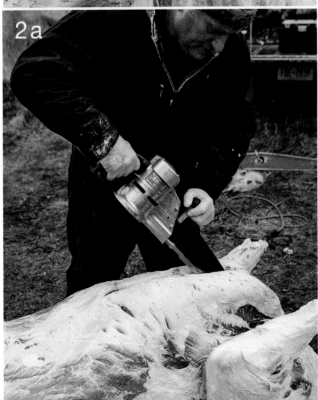

Separate the Esophagus Using a Weasand Rod

A weasand rod is a useful tool for separating the esophagus (also called the weasand) from the trachea, thus allowing you to remove the abdominal viscera and thoracic viscera separately. (The esophagus is attached to the rumen, which comes out with the abdominal viscera.) This separation can be done by hand, but the weasand rod does the job more cleanly and can separate the esophagus farther into the carcass, so that you do not have to work elbow-deep in the brisket. If you are not using a weasand rod, skip to Insert the Hanging Implement (page 108) and remember not to tie off the end of the esophagus; information on removing the esophagus without a weasand rod is covered in the section Viscera Removal without a Weasand Rod (see page 121).

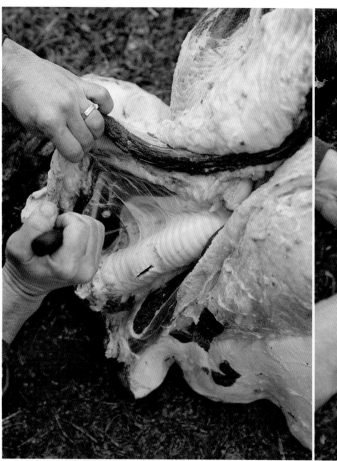 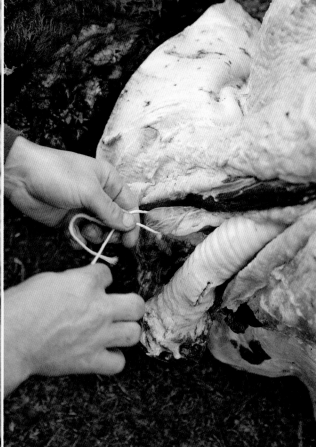

1. SEPARATE THE TRACHEA AND ESOPHAGUS.

The trachea is the larger, ridged tube — it looks like the hose part of a vacuum cleaner — and the esophagus is a soft red tube attached to the trachea. Use your knife to sever the attachments surrounding these two tubes, removing them together.

2. TIE OFF THE ESOPHAGUS.

Find the esophagus, separate the end from the trachea, and tie off the end with a piece of twine.

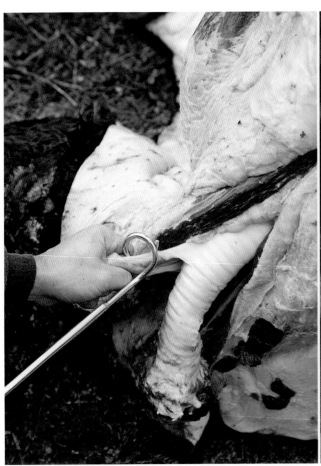 

3. THREAD THE WEASAND ROD.

Thread the esophagus through the spiraled end of the weasand rod.

4. PUSH THE ROD INTO THE CAVITY.

Hold on to the end of the esophagus that is closest to you and then push the rod into the thoracic cavity of the carcass. Be firm; this process can require a considerable amount of pressure. Make sure that you push the rod as far into the cavity as you think is possible while avoiding any tears or causing damage to the esophagus. Remove the weasand rod. If you bled the animal using the thoracic method, after separating the esophagus sever it several inches after (anterior) where you tied it.

Insert the Hanging Implement

With the gambrel cords exposed, set up for lifting the animal. Insert your hanging implement — gambrel, singletree, meat hooks, and so forth — into the space between the gambrel cord and the tibia. Hoist the animal enough to securely set the hanging implement in place.

Sever the Bung

1. HOIST THE REAR OF THE ANIMAL.
The bung should be at a comfortable working height for you, a bit below eye level. If you have to reach up to access the bung or do not have a clear view down onto it, lower the carcass.

2. SKIN THE ROUND AND RUMP.
Continue skinning the hide from the round and rump, making sure that it drapes away from the carcass. Clean away any manure that may be coming out of the bung. If you are working with a male animal, cut the pizzle loose from the abdomen, sliding your knife beneath it, and then follow the pizzle to its pelvic origin. Then, hold the pizzle straight up and taut. Keeping your knife vertical, use the tip to cut in front of (ventral to) the pizzle, staying close to the pelvic bone. Make small cuts through the connective tissue while pulling up on the pizzle until you hear it pop free.

Sever the Bung CONTINUED

The bung is the lower of the two orifices; the vagina is on top.

3. SEVER THE BUNG.

To continue removing the hide from the rump, and before you reach the tail, you will need to cut across the bung. When you reach it, pull the hide and anus away and cut straight across the entire bung. The bung will retract into the carcass a bit; this is normal. Continue removing the hide past the bung until you reach the tail. Once this procedure is complete, thoroughly wash your hands.

CAUDAL CUTTING

The bones in the tail are the caudal vertebrae. Each caudal bone is connected by a seam of cartilage that is soft enough to knife through; the challenge is in finding the seam. Fortunately there is an indicator to look for: a fine white line that can be seen, and definitely felt, on the ventral side, which is facing upward when you're removing the tail. A sharp knife should cut through the cartilage with little effort, so if you find yourself hitting bone, stop, shift your knife a bit, change the angle here and there, and try to finesse your knife through the joint. Attempting to force this cut will only dull the knife.

Remove the Tail

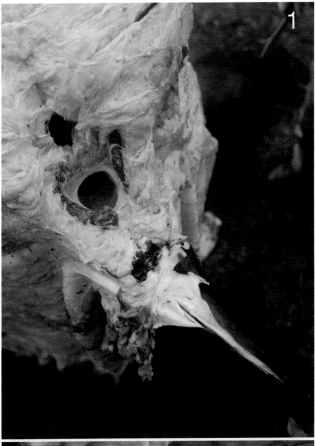

1. SCORE THE HIDE.
Run your knife under the hide for the entire length of the tail.

2. SEVER THE TAIL.
Find the base of the tail (first caudal vertebra) where it connects to the sacrum. Notice the indicator of where to cut, the cartilaginous connection between the bones. Separate the base of the tail by cutting through this connection.

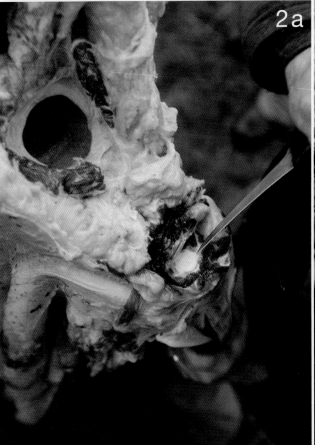

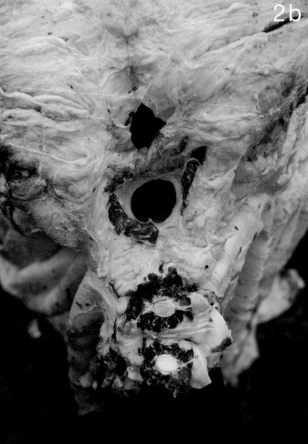

Remove the Tail CONTINUED

3. RELEASE THE TAIL FROM THE HIDE.

After you've severed the joint, cut around the base of the tail, separating it from the hide. Grab hold of the base of the tail and pull it away from the carcass to release it from its hide sheath, using your knife when necessary. Save it for oxtail.

Tie Off the Bung

After separating the anus (rectal opening) from the bung, the bung needs to be isolated and tied shut, thus preventing more feces from being expelled. The bung runs through the pelvic opening, at the center of the pelvis. The bones around the pelvic opening have changing angles: the sides tend to be around 45 degrees, whereas the front (ventral face) and back (dorsal face) are almost 90 degrees. As you cut around the bung, change the angle of your knife tip accordingly.

 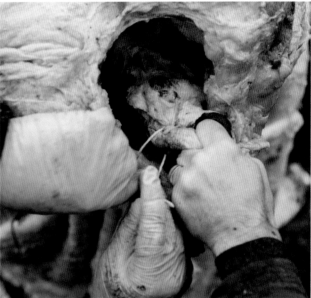

1. SEPARATE THE BUNG.
Hold the bung by the opening and start cutting along the sides of the bung using the tip of your knife. Finish by following the shape of the pelvis until you can pull the bung out far enough to tie it off.

2. TIE OFF THE BUNG.
Using a piece of strong twine or cord, tie the bung closed. If the animal is male, tie off the bung with the pizzle. Regardless, two people are needed for this process: one to hold the bung, and another to tie the bung. Alternatively, you can place a plastic bag over the end of the bung and secure it with a rubber band or string to prevent any contamination around the bung opening from getting on the carcass. Once tied, push the bung into the cavity for retrieval later during evisceration.

SPLITTING THE PUBIC SYMPHYSIS

There is one notable variation to tying off the bung that you may choose to incorporate or not. The two sides of the pelvis are joined by a weak strip of cartilage (called the *pubic symphysis*) that runs vertically right in the middle of the hind legs. This strip of cartilage can be split, giving you easy access to the bung from the front of the carcass. If you choose this method, proceed with caution: right behind the seam sit the bladder and bung, two things you don't want to damage.

Cut through the leg muscles that join at the center of the groin, following the connective tissue seam between them, until you reach bone. Notice the centerline of cartilage. Saw through the pelvis at this line, proceeding slowly so as to not damage any of the viscera. When the seam is split the pelvis will spread, allowing you to cut around the sides and back of the bung. Once separated, tie it off as you normally would before proceeding with evisceration.

SKINNING VEAL

VEAL CALVES never have the fat coverage of their older relatives. This makes aging the veal carcass a challenge, as it will often dry out quickly, creating unpalatable meat before the aging process is complete. Because of this, veal carcasses benefit from having the fell, an outer membranous layer underneath the hide, left on the carcass as a barrier to moisture loss. Therefore, when skinning veal it is advantageous to stay as close to the hide as possible to maximize the amount of fell and fat left on the carcass, allowing for more protection while aging. You can also drape the carcass with cheesecloth or caul fat to help stem moisture loss during aging.

There are two approaches to skinning veal: with a knife or with your fist (a process known as "fisting"). The process for skinning a veal carcass with a knife is much like that of beef (see page 103). Fisting involves getting between the fell and the hide with your fist, causing them to separate based on their natural seam. There are still times when you need to use a knife, but because there is minimal cutting involved, fisting can provide a cleaner hide and more presentable carcass while maximizing the amount of fell and fat left on the carcass.

FISTING IS DONE in three main phases that correspond roughly to the primals: sirloin and legs first, then the loins and belly, and finally the shoulders and foreshanks. Take note of a couple of basic tenets:

- Your hand should be not a clenched fist, but rather a tightened shape that reduces the overall surface area so that you can insert your hand into the tight spaces between the hide and the carcass. It may feel like trying to force your hand through the sleeve of a wet T-shirt.

- You will be using your second knuckles primarily. In sliding above the fell, wet hands work better than dry hands, so keep your clean hand moistened (which should not be a problem if you have remembered to wash your hands regularly during the slaughtering process). Use your dirty hand to keep the hide taut, and use your clean hand to press against the hide (not against the muscle).

- The hide should feel smooth; if it starts to feel stringy or textured, then your fist has fallen below the fell. In that case, stop fisting. Use your knife to cut through the membranes attached to the hide, working your way back to the dry look of the hide. Then continue fisting.

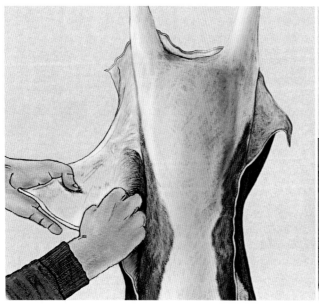
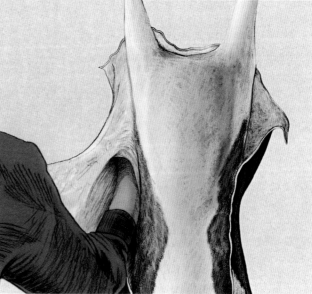

WHEN YOU PLAN on fisting a veal hide, you can begin the process the same way as you start when knifing: open up the hide at the groin and begin to release the flank area of the hide until you have a flap large enough to hold on to and have reached the side of the carcass. Then hold the flank flap of hide taut with your dirty hand and, near the base of the leg, use the thumb of your clean hand to start making a gap between the hide and the fell. (When fisting a hide, always use your thumb or a finger to create the beginning gap.) From there, slowly tunnel your fist and arm deep into the hide, leveraging your arm downward to enlarge the area and continuing to tunnel upward toward the leg.

This will get you started, and from here you can continue fisting both sides until they meet in the back, at which point pulling down on the hide until you reach the forequarter will be easier. Be careful not to tear into the back fat or muscles when pulling down on the hide; use your knife when necessary.

THE INCISION MAP for fisting is similar to that for knifing, so when in doubt follow the instructions in this chapter on where to open up the hide to help it drape away from the carcass and to make skinning the forelegs and sternum easier.

Finish Skinning

Hoist the animal and continue to remove the hide along the back, abdomen, and forelegs. Pull down on the hide to peel it away from the back, but aim to leave as much fat on the carcass as possible.

Remove the Abdominal Viscera

With the esophagus separated and the bung safely inside the cavity, and tied off to prevent any contamination, it's time to open up the abdominal cavity and remove the paunch and other viscera. The hide should be almost completely removed now, connected to the carcass only by the back of the forequarter. If you are planning to catch the viscera in a gut cart or similar vessel, for disposal or later usage, position the container now between the forelegs.

1. MAKE THE INITIAL INCISION.
Start evisceration by pulling on the abdominal wall a few inches below the pelvis and making a small horizontal incision. Once open, enlarge the opening, cutting vertically down until it is large enough to insert your hand and knife handle.

Remove the Abdominal Viscera CONTINUED

2. **SLICE OPEN THE ABDOMINAL CAVITY.** Open the abdomen using the "whole-handle method," as follows. Hold the knife handle and insert it completely into the cavity with the blade pointing out and the edge facing down. The abdominal wall should sit right at the heel of the knife, where the handle and the blade meet. With this position, you can drive the knife straight down the center of the abdomen. If you are right-handed, stand to the right side of the carcass and reach around to do this; if you are left-handed, vice versa. This is a good habit, because the weight of falling viscera can cause the knife to lunge forward mistakenly, so standing aside assures that you will not be in harm's way. Once you start cutting, you should not stop until you reach the sternum. When you stop, the guts will fall out and continuing the cut will be incredibly difficult. When you reach the split sternum, pull your knife away and let the viscera fall.

3. SEVER THE VISCERAL CONNECTIONS.

Notice the attachments that keep the viscera attached to the interior of the carcass, preventing them from falling to the ground. (Before you sever them, consider saving the caul fat, a delicious and gorgeous lacy layer of fat that surrounds most of the rumen. You can easily peel away the caul fat and save it as a gathered bunch of lacework, to be frozen and used later for wrapping roasts or rendering into tallow.) Take special care when working near the bladder and, while you remove the other viscera, avoid any actions that may cause the bladder to burst or leak while it is inside the cavity. Grab the bung and pull it out of the carcass. Start severing the visceral attachments, starting with the ureters, while rolling the paunch and intestines away from the carcass until they're released (the other viscera will follow).

COLLECT THE CAUL FAT FROM VEAL

The caul fat that surrounds the four-chambered stomach and parts of the intestines will help stem dehydration of a veal carcass while it ages (see page 133). Make sure to collect the caul fat during eviscerating, before the rumen is fully removed. Save it in a clean container for use during the cooling and aging process. You don't have to rinse it unless it has been contaminated during removal.

Remove the Abdominal Viscera CONTINUED

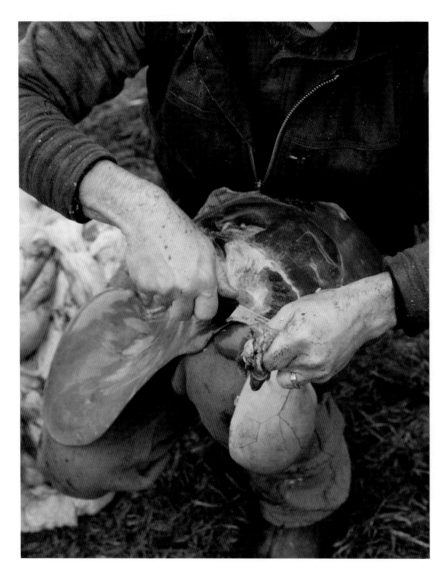

4. REMOVE THE LIVER.

(If you do not want to separate the liver for later use, skip this step and remove the liver with the rest of the viscera.) Find the liver and cut away any remaining attachments to free it from the carcass. Note the bile sac (gallbladder). Pinch the duct where the sac meets the liver; sever the connection and then carefully tear the duct off the surface of the liver, using your knife when necessary, and then discard. Trim away any other ducts or hardened areas. (Should you spill bile onto the liver, just continue removing the sac and thoroughly rinse the liver before saving.) Find the spleen (look for the purplish-red hue), which is nestled close to the paunch. The spleen is edible; save it if you so desire.

IDENTIFYING LIVER FLUKES

One of the most common internal parasites you will encounter in beef is liver flukes. There are three species of liver flukes that your beef may have: the common liver fluke (*Fasciola hepatica*), the deer fluke (*Fascioloides magna*), and the lancet fluke (*Dicrocoelium dendriticum*). Due to their reproductive cycle, which requires standing water, they are most common on farms that contain swamps or frequent flooding. But everyone should take the time to look over a beef liver after its removal for signs of damage from liver flukes.

Liver flukes migrate to the liver of mammals, where they feed until they are adults. At this point, they head to the bile ducts to lay their eggs. A sure sign of liver fluke activity is scar tissue in the liver, often viewable on the surface but maybe inside the liver itself. You may also find live or dead flukes around the bile ducts where they congregate to lay eggs. Look closely for any small, flat parasites in that area; live flukes will move when touched with the tip of a knife. Whenever you suspect damage from liver flukes, dispose of the liver and do not save any part of it for consumption by you or other animals.

5. REMOVE THE ESOPHAGUS.

If you separated the esophagus by using the weasand rod, it should easily pull through the diaphragm. In an effort to avoid tearing the esophageal walls when removing it, pull on the esophagus straight up rather than laterally. Keep in mind that the anterior end of the esophagus will be tied off and under pressure from gas, rumen contents, and gravity.

VISCERA REMOVAL WITHOUT A WEASAND ROD

When not using a weasand rod, you have two options: remove the pluck (lungs and heart) with the paunch or sever the esophagus and remove the paunch without it. Removal with the pluck is the cleaner of the two methods. To do so, follow the instructions on page 123 for pluck removal while keeping the esophagus connected to the paunch. After its removal you can come back and take out the liver and other edibles.

If you plan to sever the esophagus (if you didn't use a weasand rod and aren't removing it with the pluck), then make sure the esophagus isn't tied off at the neck. Have someone pinch the esophagus near its connection with the rumen. (Alternately, you can tie the esophagus off at this point, using sturdy twine, if you prefer.) Once tied off, sever the esophagus just below where it is pinched. Maintain the pinch to prevent rumen contents from spilling while you pull the paunch and other viscera safely out of the carcass and onto the ground or into the gut cart.

Remove the Abdominal Viscera CONTINUED

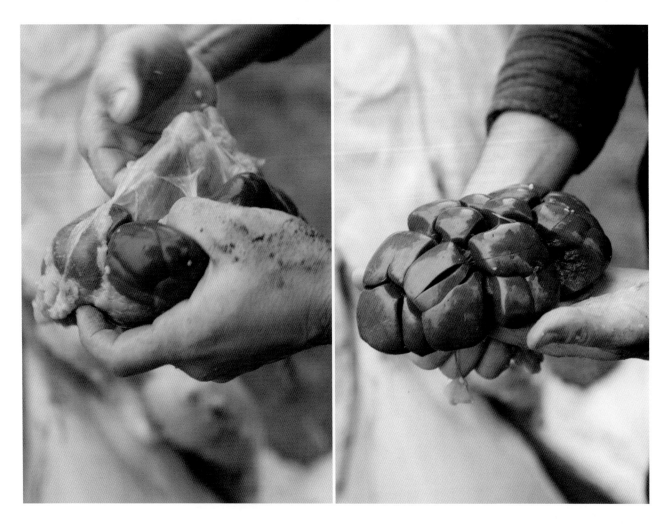

6. REMOVE THE KIDNEYS.

The kidneys of cattle are multilobed, large, and surrounded by a generous amount of fat. If you plan on cooking with the kidneys, remove them. (Otherwise, leave the kidneys with their fat in the cavity to protect the tenderloin and flank during aging.) The kidneys have a thin membrane covering that needs to be removed. Make a shallow incision with your knife from one end of the kidney to the other. Get under the membrane with your fingers and peel it away, then rinse and store the kidney.

Remove the Thoracic Viscera

You should now be able to see the diaphragm — a thin layer of muscle and connective tissue that separates the thoracic and abdominal cavities. (You may want to hoist the animal higher at this point to make working with the thoracic cavity more comfortable.) In the middle of the diaphragm locate a sheath of white fascia. Cut through the center of this fascia from one side of the cavity to the other. (The diaphragm is under tension and will retract toward the ribs when this fascia is cut.)

The thoracic viscera — lungs, heart, and trachea — are also called the pluck. In order to remove the pluck, your arm will need to be able to easily pass through the space between the sides of the split sternum. The edges of the cut bone are sharp and can cause lacerations, so spread the split sternum if the cavity space is not wide enough. Use a clean object (e.g., a stainless-steel rod) to spread the sternum, or have someone hold open the sternum while you remove the pluck.

Reach into the chest cavity, grab the pluck, and pull down and out. Exposed at the back is the thoracic aorta, a large blood vessel that keeps the pluck attached to the inside wall of the chest cavity. Sever the aorta and any other attachments to release the pluck. Cut the heart loose and remove the exterior membrane (pericardial sac). Cut across the top of the heart, leaving part of it attached, like a hinge, and squeeze it to expel any interior blood clots. Slice open the chambers and check for any remaining clots. Clean them out, and then thoroughly rinse and save the heart. Discard the lungs and trachea.

GATHERING THE SWEETBREADS FROM VEAL

The thymus gland, also known as sweetbreads, is largest when an animal is in its adolescent phase. As the animal matures past adolescence the thymus gland atrophies, or shrinks, until it is of negligible size. This gland is located behind the sternum and in front of the heart. After splitting the brisket (see page 105) and removing the thoracic viscera (see above), take the time to remove the sweetbreads if you plan to use them. Rinse and then freeze them immediately to ensure freshness (see pages 86 and 314).

Remove the Head

As with any slaughtered animal, the skull is removed where it connects to the first cervical vertebra, called the atlas joint. Disjointing is best approached from the dorsal (top) side of the neck. Vertebrae are tricky bones to disjoint, as their shapes overlap and cradle each other. They are best tackled with the tip of a sharp knife, which allows you to access the small spaces between the bones while severing the connective tissues that hold them in place.

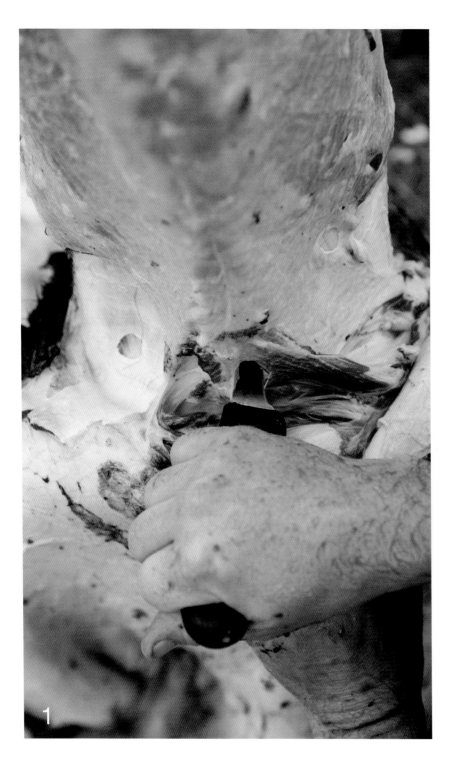

1. SEVER THE NECK MUSCLES.

Feel for the base of the skull. The atlas joint sits just above this location. You should be able to insert your knife into the middle of the joint by approaching it from the top of the spine. Find that point and then cut all the way around the neck, exposing the joint.

2. DISJOINT THE ATLAS.

Use the tip of your knife to explore the crevices within the joint, focusing on disjointing one side on the joint.

Once you have released one side, the remaining connective tissue should be easily severed. If you skinned the head and want to gather the cheeks, you won't want the head falling on the ground. Make sure someone has a good hold of it (an adult head can weigh as much as 30 pounds) either by an ear or maybe with a meat hook around the mandible (jaw bone).

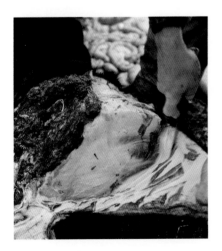

SKIN THE HEAD

If you plan on tanning or selling the hide, you should skin the head. Make an incision along the underside of the jaw. Skin one side of the face first, staying close to the hide if you plan on saving the cheeks, until you reach the flat middle of the face. Repeat on the opposite side until the head is fully skinned. Finish removing the entire hide by separating the final connection point at the neck.

Remove the Cheek Muscles

Beef cheek meat is an excellent cut for braising. This cut is certainly worth saving, although it's frequently discarded or overlooked. (If you have not already skinned the head, follow the aforementioned directions, proceeding only far enough to expose the entire cheek muscle.) Place the head on the flat center of the face, jaw facing up.

1. RELEASE THE MUSCLE.
Make an initial cut along the bottom edge of the mandible, deep enough to lift the edge of the cheek meat. Keep your blade flat and pointed slightly toward the bone. A semi-stiff blade helps here; its slight bend allows you to apply pressure and keep the blade against the bone. Continue cutting underneath the muscle and against the mandible, using the other hand to peel away the muscle, until you reach the point where the mandible inserts into the skull.

2. REMOVE THE MUSCLE.
Finish removal by cutting the muscle free at the juncture with the skull. Repeat on the opposite side of the head. Trim away the exterior fat and connective tissue.

Remove the Tongue

The bovine tongue should not be overlooked as a delicious source of meat. Think about it: if working muscles develop a greater flavor than sedentary ones, the tongue is bound to be delicious. The removal is quick and easy: a couple of quick cuts and you've got yourself a meal for the family.

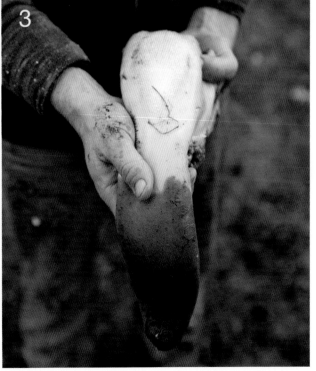

1. **CUT DOWN THE LENGTH OF THE JAW.**
Place the skull on the flat of its face, jaw up. Near the base of the skull, feel for the larynx (voice box). Insert your knife in front of (anterior to) the larynx and make an incision across the back of the tongue and between the interior edges of the jaw bone, deep enough to open the sinus cavity. You should be able to insert your fingers and feel the cavity and the bony larynx. The interior edges of the mandible will serve as your guide when you cut out the tongue. Insert your knife along one of the edges and cut all the way from your initial incision to the front of the jaw.

2. **RELEASE THE TONGUE.**
Repeat on the opposite side, thus releasing both sides of the tongue. Roll the tongue out through the bottom of the jaw and sever any remaining connections at its base.

3. Wash thoroughly and save.

Split the Carcass

Now that the carcass is fully eviscerated and the head has been removed, the carcass is ready to be split. There is a single objective for splitting a carcass: to evenly separate the two sides of the animal by cutting directly through the center of the spine. Deviation to one side or another most often happens in the lumbar and thoracic vertebrae, and a bad cut risks damaging the most valuable meat: the loins. Ideally, splitting is done with an electric saw, though it can certainly be done with a large bone saw and an ample amount of exertion.

1. START IN THE FRONT, FINISH IN THE REAR.
It's easier to see the shape of the pelvis, and choose your midline, from the front of the carcass. Hence, begin splitting while standing in front. Saw through the front (ventral) side of the spine and make your way through the pelvis and sacral vertebrae. When you hit the lumbar vertebrae, you do not have to lean into the carcass while sawing. Instead, walk around to the back to continue sawing through the spine. Have someone help steady the carcass while you saw. While standing behind the carcass, it will be helpful to have your assistant guide your work to ensure that the saw blade stays on center.

KEEP VEAL WHOLE

A veal carcass will benefit from aging whole rather than splitting, similar to that of smaller animals like sheep and goats. However, if you have space constraints, consider splitting the carcass between the fore and hind saddles rather than lengthwise along the spine. Follow the instructions for quartering the carcass on page 132, cutting between ribs 12 and 13 (knowing, of course, that you are working with a whole carcass, rather than sides of beef).

2. SAVE THE HANGER STEAK.

Before you start, note the hanging tender (also known as the hanger steak), a muscle that previously connected the diaphragm to the spine before you severed the diaphragm. Ensure that it is saved in one piece by having someone hold it to one side when the saw passes by it.

3. FINISH THE SPLIT.

Saw all the way down the spine until you reach the neck. If you are going to be moving the hanging carcass (e.g., driving it closer to a cooler, or moving it from outside to the inside of a barn where you will hang it) and it's hanging on a gambrel, singletree, or other setup that requires the two sides to be in balance, leave 6 to 12 inches of the neck intact. This connection will help keep the carcass sides balanced during transport, preventing one from slipping off the gambrel and onto the ground (or worse, landing on top of someone). If you will not be moving the carcass, saw through the neck and finish splitting the carcass.

Carcass Trimming

INEVITABLY THERE ARE SOME areas of the carcass that were contaminated during the slaughter. Take the time to trim away any obvious areas of contamination, and perform a final rinse before hanging for cooling. Make sure your hands and utensils are clean and then inspect all areas of the carcass (inside and outside) for any foreign matter, such as hair, manure, rumen contents, dirt, and so forth, or abnormalities like abscesses. The goal here is to trim as little as possible, but enough to remove the unwanted materials. Areas where the hide was opened (hocks, shanks, neck, and the midline of the abdomen) are sure to have contaminants that will be in need of trimming. The interior cavity may have remnants from the evisceration, like pieces of the liver or pizzle, or spilled rumen contents in need of rinsing.

Rinsing the carcass after trimming is a good idea only if you have good circulation in the space where the carcass will be cooling and aging. Without good circulation, the excess moisture from rinsing can cause the carcass to spoil, certainly something you want to avoid. To rinse the carcass, use a hose affixed with a clean sprayer tip, to avoid spraying any contaminants on the carcass. Do not use anything (such as a towel) to dry the carcass after rinsing; rather, allow it to dry naturally within the well-circulated environment you will use for cooling.

Quarter the Carcass

For many people, quartering the carcass while the sides are hanging will be the preferred approach: quarters are possible to maneuver by a single person and require less vertical hanging space for aging.

1. SEVER THE LOIN MUSCLES.

Find the thirteenth (last) rib. Insert your knife between the twelfth and thirteenth ribs about halfway between the loin and the flank. Cut toward the loin, severing the loin muscles down to the bone.

2. SPLIT THE RIB AND SHORT LOIN.

Saw through the spine at the point where you severed the loin muscles.

Quarter the Carcass CONTINUED

3. SPLIT THE FORE- AND HINDQUARTERS.

Secure the forequarter you are about to separate, either with someone holding it using meat hooks (as shown in the photo) or a hanging hook inserted between the eleventh and twelfth ribs. (Remember, if your setup requires balance, have the opposite carcass side supported during this step.) Once secured, follow the cut between the last two ribs, heading away from the loin, through the costal cartilage, and out through the flank to finish the separation. Remove the hindquarters by unhooking them at the gambrels, either one at a time or in unison if your setup requires balance.

Cooling, Aging, and Cleanup

THE SOONER A SPLIT CARCASS begins cooling the better. (If your hanging space is low, it may be better to cool and age the carcass in quarters. Follow the instructions in the preceding section, which covers splitting the sides of beef between the twelfth and thirteenth ribs.) Take an initial carcass temperature by inserting a sanitized thermometer down to the bone near the rump or round. This will allow you to monitor the progress of cooling. Ideally, the carcass should drop below 40°F within 24 to 36 hours after slaughter. Return in 24 hours and take another temperature reading to ensure that the process of cooling is progressing as expected.

After the carcass is cooled, it will need to be aged. The details of aging are covered on page 18, but at minimum beef should be aged for 10 days, while ideal aging is 14 to 21 days.

All offal that isn't going to be used immediately benefits from being frozen soon after slaughter, but first it will need some processing prior to storage. Details for cleaning and processing offal are found on page 85.

Cleaning up the waste from slaughtered cattle requires some carting and heavy lifting, so preparations should be made in advance. Typically, cattle viscera are composted due to the considerable size and weight — tossing a 200-pound rumen in with the trash isn't a great option and, unless you are comfortable attracting scavengers, carting the entrails into the woods isn't an ideal disposal solution either. If necessary, the weight and volume of the viscera can be greatly reduced by cutting open the rumen and emptying its contents. Hides in relatively good condition can be sold or tanned. Recommendations and further information on cleanup can be found on page 89.

COOLING AND AGING VEAL

Along with achieving the desired effects of aging, the main priority when hanging veal is to prevent excessive dehydration of the meat. Without the fat cover that is typical on beef, a veal carcass can lose moisture at an alarming rate. Thus, it is imperative to keep a high humidity — around 80 percent — where you are hanging the carcass. In addition to the guidelines on page 87 about cooling and aging, plan on draping the animal's caul fat over the rounds and rump, and then wrapping the carcass in cheesecloth to further lessen uncontrolled moisture loss.

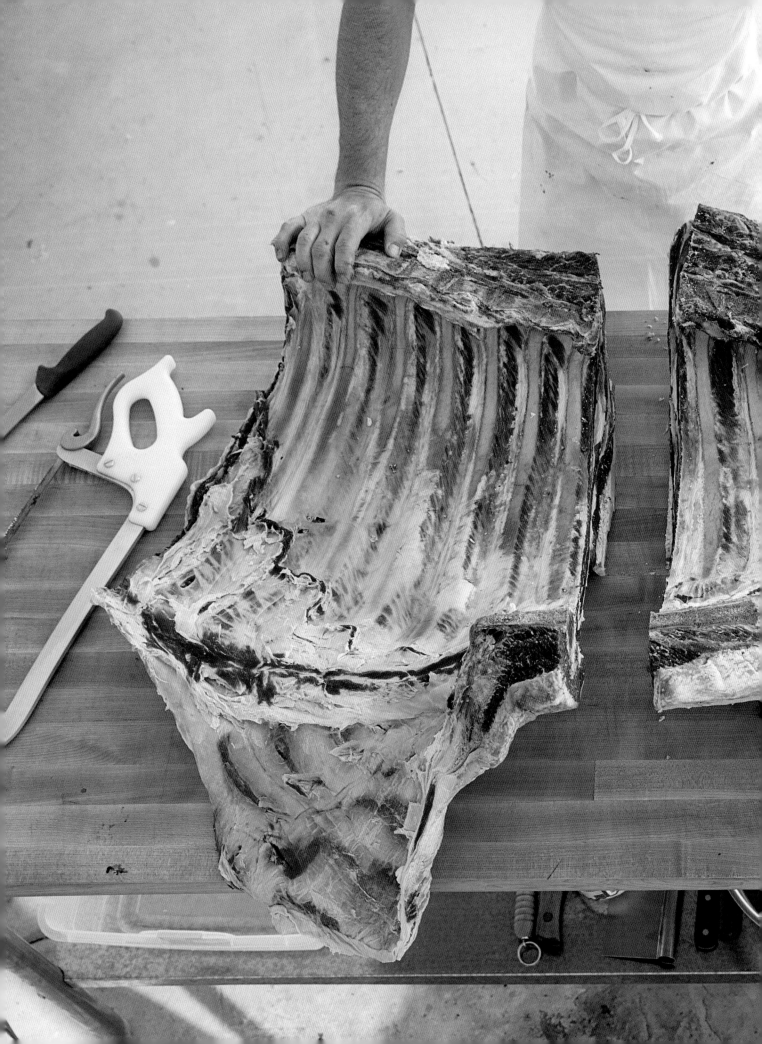

BEEF BUTCHERING

BUTCHERING AN ENTIRE SIDE OF BEEF is a learned skill. Learning to butcher beef is worth the effort because this skill is transferrable to working on any common livestock animal. There are many ways to break down a side of beef, and this chapter details several variations. After working through a side of beef using the instructions provided here, you can build on that knowledge to develop a system that works best for you. Your individual system can be based on your own culinary habits, taste preferences, and cultural traditions. After you have established your own basic butchering skills, you may want to expand your repertoire to include butchering traditions from other places around the world.

Butchering Setup, Equipment, and Packaging

THE PRIMARY CHALLENGE when accommodating beef butchering is space: you will do most, or all, of the work on a table. You'll need a table that is at least 2 feet by 3 feet. This size is a bare minimum; you'll feel cramped and you won't have much room to maneuver, but you can get the job done on a table of this size. A more manageable table size is 3 feet by 5 feet, or larger. Regardless of the table's size, it must have a top that can be easily sanitized and that is dense enough to withstand cuts from your knife and saw. Some commercial tables feature steel structures and removable plastic cutting boards. The removable cutting boards make cleaning relatively easy, especially if you have access to a large utility sink that can accommodate individual boards. A traditional wood-top butcher block — my personal preference — performs very well, and can be cleaned and sanitized in place. (Follow the sanitization instructions on page 30 for either surface.)

Equipment

Butchering a whole beef carcass into portioned cuts requires a few essential hand tools and pieces of equipment. A wide array of optional items can aid in specific tasks. Do not look for prized brands or expensive knives. Instead, buy a knife that will hold an edge when honed, has a handle that can be easily sanitized, and is affordable. In addition to the required tools, always have paper or fabric towels on hand as well as butcher's twine. Detailed descriptions and recommendations for tools and equipment are found in chapter 2.

REQUIRED EQUIPMENT

- Bins or lugs
- Boning hook
- Boning knife
- Breaking knife
- Handsaw/bone saw
- Honing rod/knife steel

OPTIONAL EQUIPMENT

- Bone dust scraper
- Butcher knife
- Cleaver
- Mallet (rubber or wooden)
- Paring knife

PACKAGING

The methods of packaging meat are specific to how the meat is going to be stored: fresh or frozen. With beef, the assumption is that you, or your friends and family, will be freezing a good portion of the meat harvested from the carcass. Therefore, one of your main priorities is to choose a method of packaging that will stave off freezer burn. Meat that is successfully packed for the freezer is wrapped skintight and does not have any pockets of air. In most cases a vacuum sealer will provide you with the best results. For complete details on the equipment and methods for packaging and freezing, refer to chapter 8.

Primals and Subprimals

BUTCHERING BEEF IS A SUBSTANTIAL endeavor. It is physically demanding, of course, but it is also mentally challenging, as the myriad options can be overwhelming. Fortunately, this singularly large job can be broken up into smaller tasks, because a beef carcass can be separated into subsections, called primals. Each primal is a grouping of muscles that is associated with their functional similarity: the chuck contains the shoulder muscles, the loins are composed of structural back muscles, the round is made up of mostly lean and large leg muscles, and so forth. The area of each primal is defined by the skeletal structure of the animal, so that you separate one area from the other based on the location of certain bones. The result of the separation is shown on the beef meat map, which is equivalent to how a country is broken down into states. In some cases primals are further sectioned into subprimals (like when states are separated into counties). This chapter will cover the separation of main primals and subprimals, and how they are broken down into individual cuts.

One of the challenges of meat cutting is the local vernacular — the label for any single cut of meat may change from country to country, and even regionally within each country. In an attempt to help alleviate some of the confusion, I use the most common names of cuts for the United States, and provide step-by-step instructions for how to fabricate those particular cuts. Where and how to make cuts also varies according to cultural and regional traditions; so to simplify this information, this book will provide cutting guidelines that are largely based on those of the North American Meat Processors Association (NAMP). Please note that scientific muscle names are provided in italics when applicable, most commonly when cuts are composed of a single muscle.

BEEF PRIMALS

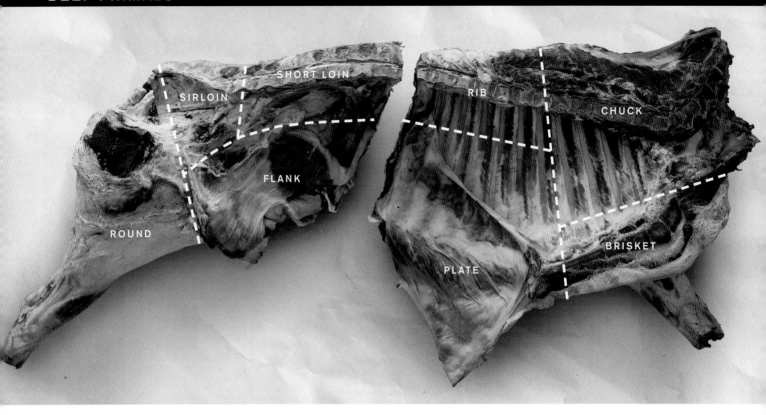

THE CHUCK is the largest primal and contains the widest variety of bones. It is separated from the rib primal with a cut between the fifth and sixth ribs, leaving the first five (vertebral) ribs with the chuck. A portion of the thoracic and the entire cervical vertebrae represent the chuck portion of the spinal column. The scapula — also known as the paddle bone — and the humerus represent the upper portion of the foreleg.

THE SQUARE-CUT CHUCK is separated from the foreshank and brisket with a cut starting near the humerus–radius joint and continuing across the rib cage. This separation leaves the sternum and sternal portion of ribs 1 through 5 with the brisket. The foreshank includes the entire radius, ulna, and, depending on the location of said cut, any remnants of the humerus.

THE RIB AND PLATE PRIMALS span ribs 6 through 12. (Ribs 1 through 5 are on the chuck primal, and the last rib, 13, is with the loin primal.) The rib primal includes the vertebral portion of said ribs and the corresponding thoracic vertebrae. A portion of the scapula, and the cartilaginous tip, are left with the rib primal after it is split from the chuck. The plate primal is separated from the rib primal, leaving the remaining vertebral portion and the entirety of the sternal portion of ribs 6 through 12.

THE LOIN PRIMAL includes the last rib and thoracic vertebra (T13), and the entirety of the lumbar and sacral vertebrae. The posterior end of the loin includes a portion of the pelvis which, after splitting, is known as the aitchbone. Depending on where the loin and round primals are split, the aitchbone usually contains a portion of the pubis and the entire ilium bone. The loin primal has two subprimals, the short loin and sirloin. These are separated at the anterior tip of the ilium, a point that roughly falls in line with a gap between the last and second-to-last lumbar vertebrae — L5, L6.

THE FLANK, the first primal removed from the hindquarter, is usually considered boneless, though it may contain a remnant of the thirteenth rib.

THE ROUND PRIMAL is the entire hind leg. It is separated from the loin with a cut through the pelvic girdle near the sacral vertebrae, leaving a portion of the pelvis — the ischium and the pubis — with the round. (This portion of the pelvic girdle is commonly referred to as the aitchbone.) The largest bone in the body of the animal is the femur, and it connects to the pelvic girdle at a ball-and-socket joint. At its distal end the femur connects to the tibia, creating the stifle joint (the equivalent of the human knee) where the patella (kneecap) sits. The tibia has, as is common with so many quadrupeds, a fused fibula. At the distal end of the tibia are a collection of fused bones (the remnants of the cannon bone and hoof that are removed during slaughter) that are commonly referred to as hock bones, the most prominent of which is the calcaneus, the bone connection of the gambrel tendon.

BUTCHERING CUT POINTS

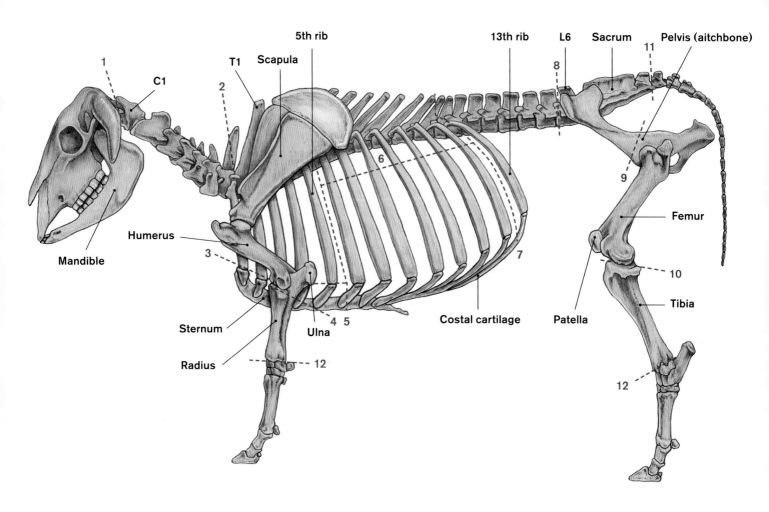

Understanding the structure and shape of the bones is key to efficient processing. Although it is beneficial to know the scientific names of bones, it is more important to understand how the bones fit together and between which ones to cut.

1. Remove head at the atlas joint
2. Remove neck at the last cervical vertebra (C7)
3. Separate the brisket from the chuck
4. Remove the foreshank at the humerus/ulna joint
5. Split the chuck and rib primals (between the 5th and 6th rib)
6. Separate the plate primal from the rib
7. Separate the fore and hindquarter (between the 12th and 13th rib)
8. Split the loin primal into the short loin and sirloin
9. Separate the round primal from the sirloin
10. Remove the hind shank at the stifle joint
11. Remove oxtail (during slaughter) at the first caudal vertebra
12. Remove cannon bones from shanks (during slaughter)

FOREQUARTER

A BEEF SIDE IS SEPARATED into two unequal halves — the forequarter and the hindquarter. The forequarter includes the neck, shoulder, forelimb, and ribs of the animal. It contains the widest array of muscles and the greatest diversity of cuts. The forequarter is sectioned into the following primals: chuck, brisket/foreshank, rib, and plate.

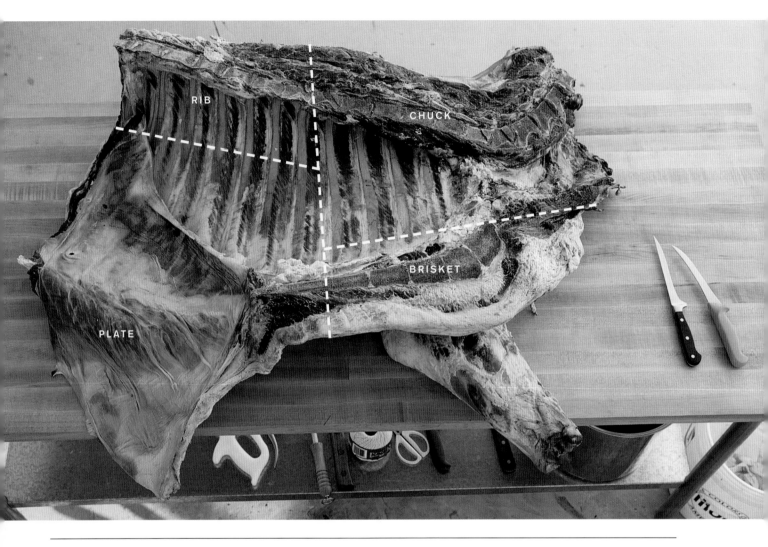

UTILIZING GRAVITY AND TENSION

Working with meat while it is hanging, or "on the hook," aids in the initial carcass breakdown. Halves, quarters, or even primals are cumbersome, and moving them onto the table to access all sides of the meat takes considerable effort. It is far easier to walk around a large, hanging piece of meat during the initial breakdown. You can approach it from many different angles, and let gravity be your aid. Severing the right connection points, combined with gravity and some tugging, will make for a speedy dissection of primals into table-ready subprimals. The chuck and round are especially suited for work on the hook. So, if you have the space, consider rigging up a hanging apparatus to hang quarters or primals in or near your workspace.

You can also utilize the benefits of tension by customizing your table a bit. On the underside of the table, away from where you stand, install a chain with links that are large enough to insert a meat hook. This chain doesn't need to be very long, just long enough to lie atop the table so that you can hook into a primal, such as a beef chuck. This setup will allow you to create a point of tension: with the primal hooked against the table, and you not needing to secure it yourself, you are free to pull muscles and maneuver much more freely.

Chuck Primal

The chuck primal is the shoulder of the animal and is the largest beef primal. Muscles in the shoulder enable much of the animal's locomotion, and as such these muscles are referred to as working-muscle groups. Because the animal frequently uses these muscles, they often contain large amounts of collagen and connective tissue. It's easy to understand, therefore, why the chuck is considered to be the primal with the toughest texture (but also, perhaps, the deepest flavor). However, the toughness of this primal belies the fact that it actually contains six of the seven most tender muscles in the whole animal. Because of the variety of muscle groups found in the chuck, quite a number of different cuts originate from it.

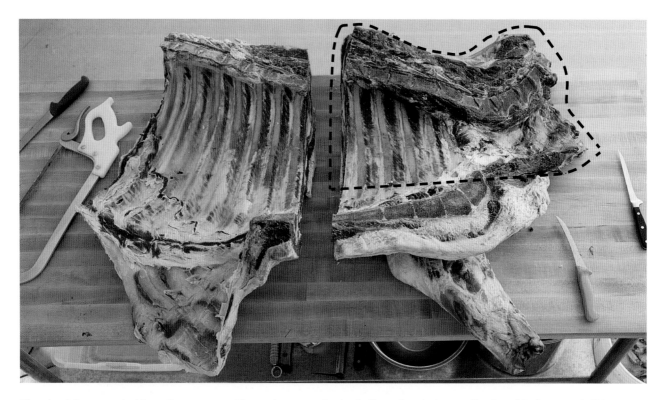

The chuck is separated from the carcass with a cut, perpendicular to the spine, between ribs 5 and 6. Commonly it is processed as a square-cut chuck, in which the brisket and shank are removed.

CHUCK SUBPRIMALS

CHUCK ROLL. The chuck roll is the largest subprimal and is, most often, a boneless cut that is made from a square-cut chuck that has been boned out after neck removal. The muscle groups in the chuck roll are those that lie beneath (medial to) the scapula. It represents the origin of many muscle groups and the termination of the coveted *longissimus dorsi* – known better as the eye of ribeye and strip steaks. The chuck roll can be cut perpendicular to its length to form individual roasts, or it can be seamed and trimmed to prepare other well-known cuts, such as the chuck eye roll or the under blade roast.

SHOULDER CLOD. Situated on the top side of the scapula, just behind its ridge, is the second subprimal, the shoulder clod. Decidedly smaller than the chuck roll, it consists of three main muscle groups that can all be seamed to form separate cuts – top blade, shoulder tender, and clod heart – and other small supporting muscles.

COMMON CHUCK CUTS: chuck roast, chuck arm roast, 7-bone roast, blade roast, chuck eye roll, Delmonico steak, top blade, flat iron, shoulder tender, petit tender, ranch steak, chuck stew, chuck kabobs, chuck short ribs, ground chuck

Brisket/Foreshank Primal

The brisket/foreshank primal consists of the pectoral muscles and the lower portion of the foreleg. The brisket is commonly available as a whole, or is seamed into two individual muscles: the larger, leaner flat cut and the smaller, more marbled point cut. This coarsely grained cut is best prepared with slow-cooking or curing methods.

Shank meat is composed of hearty working muscles that are intertwined with copious amounts of collagen that support postural function. Shank meat is most commonly found in a cross-cut form, to create more manageable pieces. This cut also exposes the unctuous bone marrow. Cooking shank meat necessitates an environment that excels at melting collagen into gelatin. A moist heat and extended cook time, like stewing or braising, is a preferred method. Both brisket and shank meat add deep flavor to ground beef mixes.

COMMON BRISKET/FORESHANK CUTS: brisket, brisket flat, brisket point, cross-cut shanks, deckle

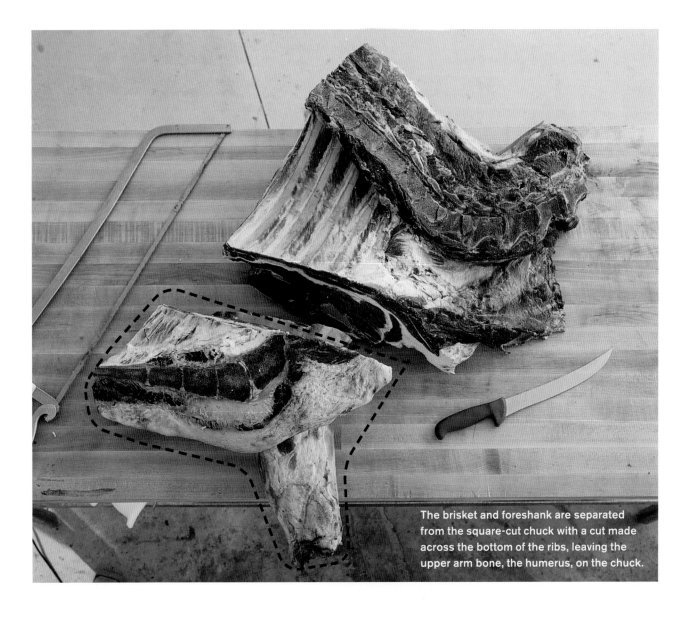

The brisket and foreshank are separated from the square-cut chuck with a cut made across the bottom of the ribs, leaving the upper arm bone, the humerus, on the chuck.

Rib Primal

The rib is one of the "middle meat" primals. This term refers to cuts from the midsection of the animal, and those which garner the highest prices. When properly aged, the rib is prized for a balance of flavor and tenderness, and exhibits deep beefy and often nutty characteristics. The rib is separated from the chuck with a straight cut between ribs 5 and 6, and is split from the short loin by a cut that follows the natural curvature between ribs 12 and 13, the second-to-last and last ribs, respectively. (If you do the math, this leaves seven ribs, 6 through 12, on the rib primal.)

Determining where to separate the rib from the plate primal may prove to be a personal preference that takes into account factors like taste, extended dry-aging considerations, and steak tail length (length of the rib bone on a steak). Commonly, the cut is made approximately 4 inches below the main loin muscle.

The main muscle in the rib primal is the *longissimus dorsi*. Because this is a structural muscle, rather than one used for mobility, it possesses a desirable tenderness. This muscle originates at the pelvis, narrows as it travels down the spine, and then terminates in the chuck. This tapering provides the drastic difference in appearance between the chuck end of the rib and the loin end. The rib primal is also known for its pervasive marbling, with the loin end being one of the inspection points that the USDA uses to determine beef grading. The cartilaginous tip of the scapula, along with muscles on the top and bottom, reside in the chuck end of the rib primal. The two muscles that are often seamed for individual cuts are the eye of ribeye and the ribeye cap.

COMMON RIB CUTS: rib steak, Delmonico steak, standing rib roast, prime rib, boneless rib roast, eye of ribeye, ribeye cap, back ribs

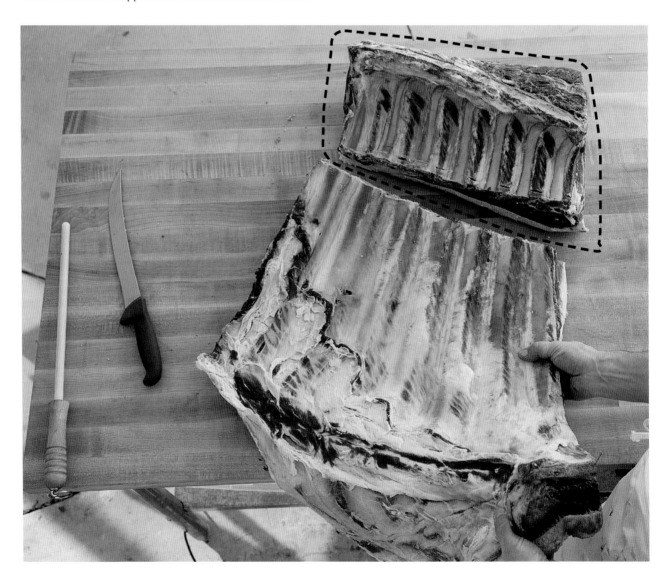

Plate Primal

The plate primal is the main source of short ribs. Short ribs are separated from the rib primal with a cut that is made parallel to the spine and several inches (or more, if you prefer) from the loin eye. Short ribs can be cut boneless or bone-in. In addition, various ethnic cutting techniques that involve additional processing, such as butterflying, can be made. Attached to the interior side of the plate is the diaphragm, referred to as the outside skirt steak. A portion of the inside skirt steak sits on the bottom edge of the plate primal.

COMMON PLATE CUTS: short ribs, boneless short ribs, flanken ribs, English ribs, royal short ribs, rib fingers, outside skirt steak, inside skirt steak, navel, Korean-style ribs

HINDQUARTER

AFTER THE FOREQUARTER IS REMOVED, the remaining portion of a beef side is known as the hindquarter. It includes the lower (lumbar) spine, flank (or belly), and rear leg of the animal. The primals are loin, flank, and round, respectively. Aside from the flank, these areas of the carcass are more infrequently worked than those in the forequarter, making for leaner, and in many cases more tender, muscles.

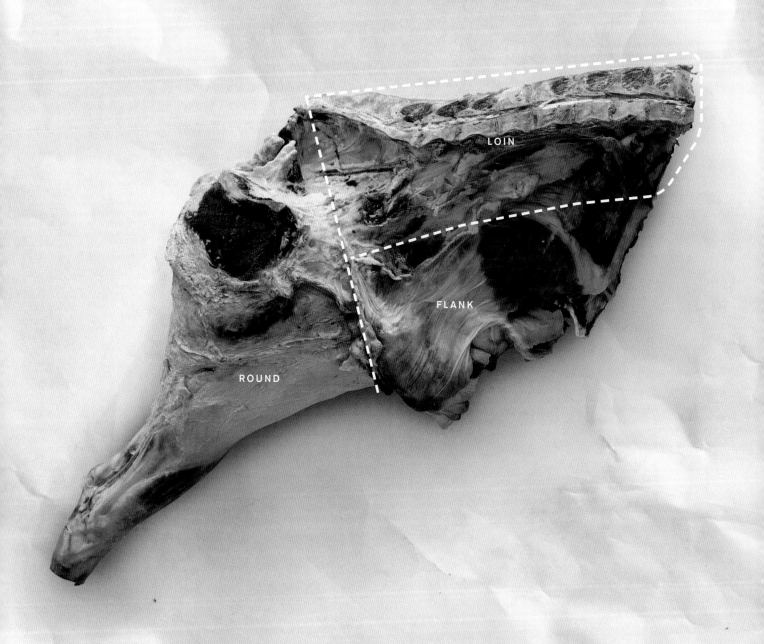

Loin Primal

The loin primal, the second of the two "middle meat" primals, is the spinal portion of the hindquarter. The front (anterior) end is separated from the rib primal with a cut that follows the natural curvature between the twelfth and thirteenth ribs; the rear (posterior) end is separated from the round primal with a cut through the ball joint of the femur. The loin is split from the flank primal by following the curvature of the sirloin flap. When making this cut, remember to leave several inches of tail on the loin eye. The loin is the most valuable primal due to its weight, volume, and quality. The loin contains a variety of muscles that are highly regarded for their tenderness and quintessential beef flavor. The most famous beef cut, the tenderloin, also known as filet mignon, sits underneath the spine and is so infrequently used by the animal that it is the most tender muscle in the entire carcass. The *longissimus dorsi,* the most valuable single beef muscle, is at its widest in the loin. This muscle continues toward the chuck, tapering until it terminates as a small muscle in the neck.

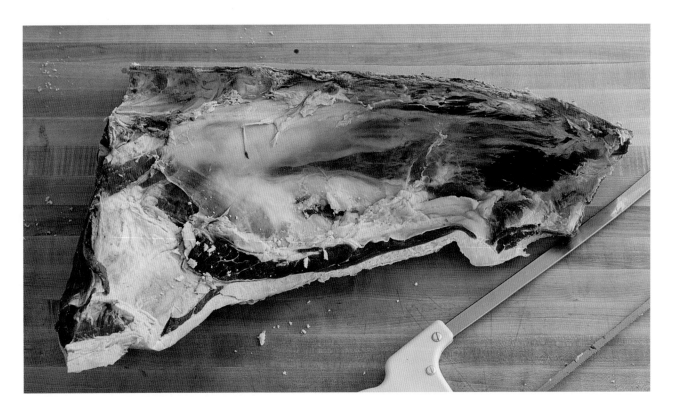

LOIN SUBPRIMALS

The full loin is composed entirely of two prominent subprimals — the short loin and the sirloin — and is split at the front (anterior) tip of the pelvis with a cut that is made perpendicular to the spine.

SHORT LOIN. The short loin is primarily made up of lumbar vertebrae. It is these vertebrae that separate the two popular muscles in the short loin — the strip steak and the tenderloin — and when cross-cut the bones mimic the shape of a "T." As you may have guessed, T-bone and porterhouse steaks come from the short loin — a piece of strip steak and tenderloin separated by a slice of lumbar vertebrae.

SIRLOIN. The sirloin is esteemed for its balance of flavor depth and tenderness. (Characteristically these two traits are inversely related, since tenderness stems largely from inactivity and flavor development occurs more within working muscles.) The three main cuts found in the sirloin are the top sirloin center, the top sirloin cap, and the tri-tip.

COMMON LOIN CUTS: strip steak, shell steak, porterhouse steak, strip loin, loin roast, boneless loin roast, tenderloin, filet mignon, tenderloin tails, sirloin steak, tri-tip, top sirloin, top sirloin butt, top sirloin cap, ball tip, rump steak, rump roast, hanging tender, hanger steak, baseball steak

Flank Primal

The flank primal is the abdominal wall, and its weight-bearing job results in musculature that is thin and strong. This beefy brawn comes from an abundance of connective tissue. The working status of these muscles, and their adjacent location to organs, contributes to the flank's unique flavor. Additionally, the flank muscles' abdominal function gives the muscle a distinct grain. Initially popularized in the 1980s as a fajita meat, the flank muscles are becoming more recognized for their individual qualities. The three cuts originating from the flank are the sirloin flap, flank steak, and inside skirt steak (not to be confused with the outside skirt steak, or diaphragm).

COMMON FLANK CUTS: sirloin flap, bavette, flap meat, flank steak, inside skirt steak, fajita or stir-fry meat

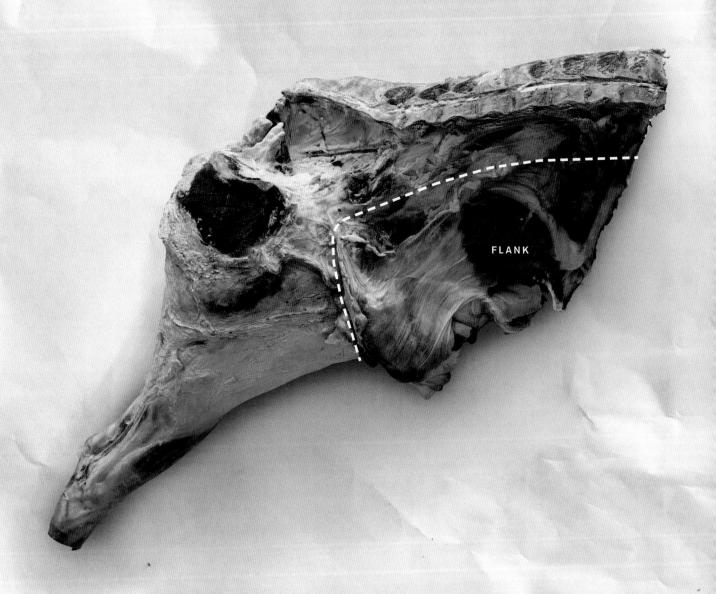

Round Primal

The final primal, the round, is the entire hind leg and, being about a quarter of the carcass weight, is the second-largest primal. The round is separated from the sirloin end of the full loin by sawing through the aitchbone — an industry term that refers to a portion of the pelvis — exposing the ball of the femur. The muscles in the round are large and lean. They are all working muscles, but they literally follow the lead of the forelegs and therefore do not require the same amount of endurance as do the muscles in the chuck. Hence, the round contains far less amounts of connective tissue, which in turn lends to a variety of preparation options.

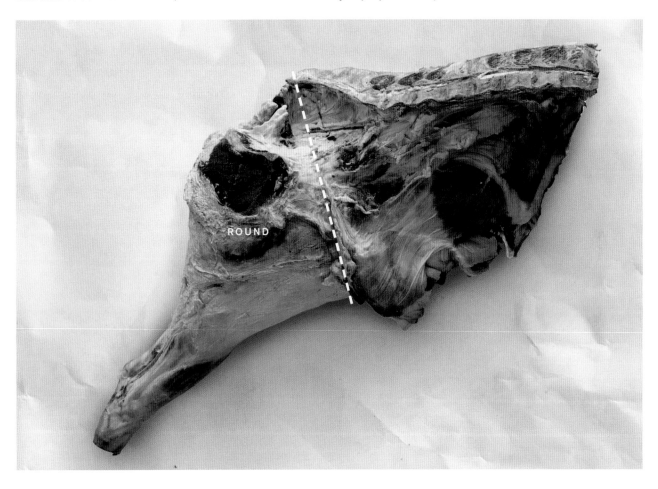

SUBPRIMALS

SIRLOIN TIP. The muscles of the sirloin tip originate in the sirloin subprimal but primarily exist in the round. The sirloin tip is composed of four main muscles, and is the bovine equivalent of quadriceps. Often the sirloin tip is seamed out for individual muscle roasts.

TOP ROUND. The top round, also known as the inside round, starts at the interior edge of the sirloin tip and wraps around to the back of the leg. Top round is the most tender round subprimal. Its primary muscle is the *semimembranosus*, the largest single muscle in a beef carcass and one of the three hamstring muscles. It is the most popular choice for lean steaks, including modern versions of London broil and minute steaks.

BOTTOM ROUND. The bottom round, also known as the gooseneck, is composed of the outside round, eye of round, and the heel. These exterior-facing muscles are in frequent motion, especially the heel, and therefore contain a considerable amount of connective tissue as compared to the other round subprimals. Preparations frequently include pickling or slow cooking, like pastrami or rump roast.

COMMON ROUND CUTS: sirloin tip roast, ball tip roast, sirloin tip steak, top round steak, London broil, minute steaks, cube steak, round roast, rump roast, Manhattan steak, eye round steak, heel side, merlot cut, western tip, western griller, Santa Fe cut, San Antonio steak, Tucson cut

VEAL

Veal — the younger, leaner, milder-tasting version of beef — can be broken down in many of the same ways that beef is. Because the veal calf is much smaller, the veal cuts will be dramatically smaller than their beef counterparts. For example, the shoulder tender of beef is, on average, just over a pound; so considering that it's a negligible cut, it therefore probably would not be separated. For veal, use your judgment to determine the size and volume of cuts that you want.

BREAST AND FORESHANK. Between the veal and beef processing maps, the separation of a breast is the main distinguishing feature. With veal, this map includes a portion of the entire rib cage as well as the sternum and the cartilage that connects the two. The cut across the arm bone and chuck portion of the rib cage leaves the foreshank with the breast. The breast is often removed before the fore- and hindquarters are separated, which allows the flank to be attached because the single-muscle cuts from the flank can often be too small for individual use.

COMMON BREAST / FORESHANK CUTS: tied and rolled breast, stuffed breast, braised breast, boneless breast, cross-cut shanks (osso buco)

CHUCK. The chuck of veal is most often fabricated as a square-cut primal after the breast and foreshank is removed. The locomotive muscles in the shoulder, and the copious deposits of collagen, make them ideal cuts for stewing and braising. The chuck is separated from the rack by a straight cut between the fourth and fifth ribs.

COMMON CHUCK CUTS: chuck eye roll, chuck roast, flat iron, chuck tender, under blade roast, chuck ribs, chuck stew, ground chuck, cross-cut shank

RACK. The veal rack, also known as the hotel rack, is one of the two middle-meat primals, spanning the fifth through eleventh ribs. The veal rack is commonly separated from the loin primal by a cut between the eleventh and twelfth ribs, making for a longer loin than that of beef.

COMMON RACK CUTS: veal chops, rib roast, frenched veal rack, boneless ribeye roast

LOIN. Veal loin, also known as the saddle, is the second of the two middle-meat primals. Together they make up the bulk of retail value for the veal carcass. After separation from the rack, the loin is separated from the leg primal with a straight cut in front of the pelvic bones, leaving the sirloin subprimal with the leg.

COMMON LOIN CUTS: loin chops, boneless loin roast, tenderloin

LEG. The veal leg primal includes the sirloin, hind leg, and hind shank. This primal is often separated into subprimals, which allows the hind leg to be prepared boneless or seamed into the typical hind leg muscle groups: top round, bottom round, and sirloin tip. The large, lean muscles of the hind leg lend themselves to the dry-heat preparations, like cutlets and leg roasts.

COMMON LEG CUTS: sirloin chops, sirloin roast, butt tenderloin, cutlets, tip roast, kabobs, cross-cut shank (osso buco)

BREAST: FURTHER PROCESSING

The breast mostly comprises hardworking, thin muscles that are woven with considerable amounts of connective tissue and help to support the animal's internal viscera. The one notable exception may be the main muscle that extends across the first several ribs, called the *serratus ventralis*. This muscle is surprisingly tender, especially compared to the surrounding rib muscles. The other muscles of the veal breasts include those of the beef brisket and plate. Depending on how the breast was removed, the flank may also be attached.

TRIMMING AND PROCESSING. The most common form of veal breast is rolled, either bone-in or boned, or with a pocket. The breast can be trimmed up before additional processing by removing the diaphragm and the thin exterior muscles (*cutaneous trunci*) that lay atop the fat. Save any clean trim for grind. If you leave the diaphragm attached, remove the exterior membrane to make it more palatable.

Bone the breast according to the directions for rib removal on page 176. To roll a bone-in breast you will need to remove the sternum and sever the costal cartilage, which allows the ribs to lie flat. The cartilage of young animals can usually be cut with a knife, but you may need to use a cleaver on the more ossified cartilage found in older animals.

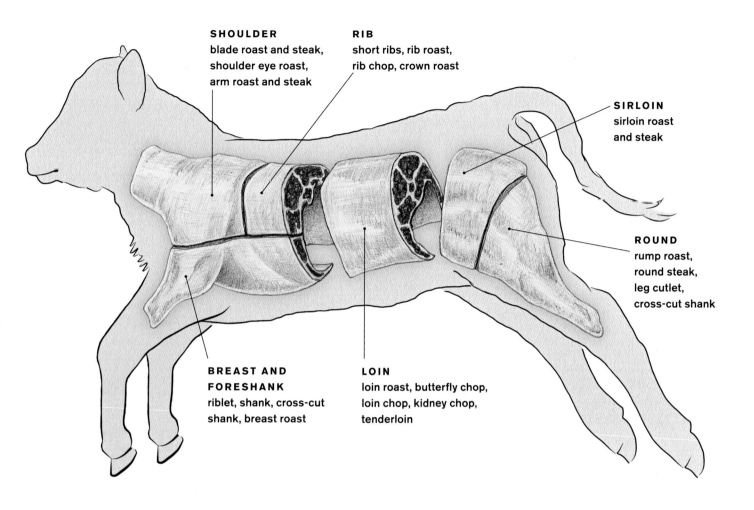

SPLITTING A VEAL CARCASS

Ideally the veal carcass has been aged whole, but often space limitations require a different approach, such as splitting the carcass into fore- and hindsaddles. Here we show you both methods: splitting the carcass lengthwise and then into quarters, and splitting the carcass into saddles.

SPLITTING LENGTHWISE. If you're able to hang and age the carcass whole, you'll need to split it lengthwise prior to processing. Follow the instructions on page 128.

FLANK SEPARATION. When splitting into fore and hind sections (saddles or quarters, whether or not you've already split it lengthwise), you must decide whether to include the flank with the breast or leave it attached to the loin. Leaving it attached to the breast gives you the option of including it in boneless preparations; otherwise, the flank is most likely destined for the grind pile.

SPLITTING INTO FORESADDLE AND HINDSADDLE. If you have split the carcass down the spine, this process will provide you with forequarters and hindquarters. If you are aging the carcass in saddles, follow the period of aging by splitting the saddles down the spine. Make your separating cut between the eleventh and twelfth ribs by counting two ribs back from the last (thirteenth) rib.

REMOVING THE NECK. Before splitting a foresaddle, you have the option of removing the neck as a whole. (A whole neck allows for bone-in neck slices or a boneless neck roast, which can then be stuffed and tied, among other preparations.) Follow the instructions on page 124.

ONE CARCASS, TWO APPROACHES

There are many different ways to break down a side of beef, and making certain cuts prevents others. For example, from a single side you can't get both porterhouse steaks and a full tenderloin roast. The following table describes two different approaches for breaking down an entire beef carcass, with each side employing a different approach. Not only does this two-sided approach provide a wide range of final-product options, it also shows the versatility of a single beef carcass.

Ideally, you should prepare a cut sheet before proceeding with any large-scale butchering endeavor. Information on cut sheets and how to make one can be found on page 68.

	SIDE 1	SIDE 2
NECK	Bone and roll; save bones for stock	Cross-cut for bone-in neck pot roast
CHUCK	Prepare square-cut and remove neck; cut thick arm roasts; cut thick 7-bone/chuck roasts until after the scapula; bone remainder for grind along with trim	Separate arm; remove chuck tender, top blade, under blade flap, and shoulder tender; remove shoulder clod, denude, and cut into ranch steaks; bone remainder, including foreshank, and use bone for marrow
		Remove neck and then bone remaining chuck; separate brisket; separate under blade roast and save for either whole or for chuck stew; cut two Delmonico steaks from rib end of chuck eye roll; tie remaining chuck eye roll into roast
BRISKET/FORESHANK	Bone brisket whole; cross-cut foreshank	Trim brisket and separate into point and flat (foreshank boned with arm)
RIB	Cut three ribeye steaks from the chuck end; prepare bone-in / bone-out standing rib roast with remainder	Remove rib cap and use for grind; separate entire rib cap and save as roast or cut into portions; cut two to three eye of ribeye steaks from chuck end and french the bones; french remaining and prepare as standing eye of ribeye roast
PLATE	Remove inside skirt; cross-cut short ribs starting at rib end until reaching the navel; bone the navel and roll	Remove inside skirt; cut flanken-style short ribs from rib end; follow with English-style short ribs until reaching the navel; bone navel
FLANK	Separate sirloin flap, flank, and portion of inside skirt; grind remaining muscles	Separate sirloin flap, flank, and portion of inside skirt; grind remaining muscles
SHORT LOIN	Cut T-bone and porterhouse steaks the entire way	Remove full tenderloin before separating sirloin and use for roast or portion; cut strip steaks from rib end; bone remainder for boneless strip loin roast
SIRLOIN	Cross-cut two to three large-format sirloin steaks starting at the round end; bone remainder for ground sirloin	Bone, then separate tri-tip, ball tip, and sirloin cap; trim remainder and cross-cut for center-cut sirloin steaks
ROUND	Separate into subprimals; denude and trim sirloin tip, prepare as roast; cut top round into Santa Fe, San Antonio, and Tucson cuts; separate eye of round for roast beef; separate heel and prepare merlot cut; prepare western cuts from outside round	Separate into subprimals; denude and seam sirloin tip for small roasts; remove top round cap, trim for Santa Fe cut; cross-cut remaining top round into London broil; separate eye of round, cross-cut, and tenderize for steaks; denude and clean outside round and roast whole; trim heel for merlot cut
HINDSHANK	Bone for grind; use bone for marrow	Cross-cut

BREAKING DOWN THE FOREQUARTER

THE FOREQUARTER INCLUDES the chuck, brisket, rib, and plate primals. When you follow natural seams, the chuck can be the most challenging and interesting primal to break down because it includes the broadest muscle diversity.

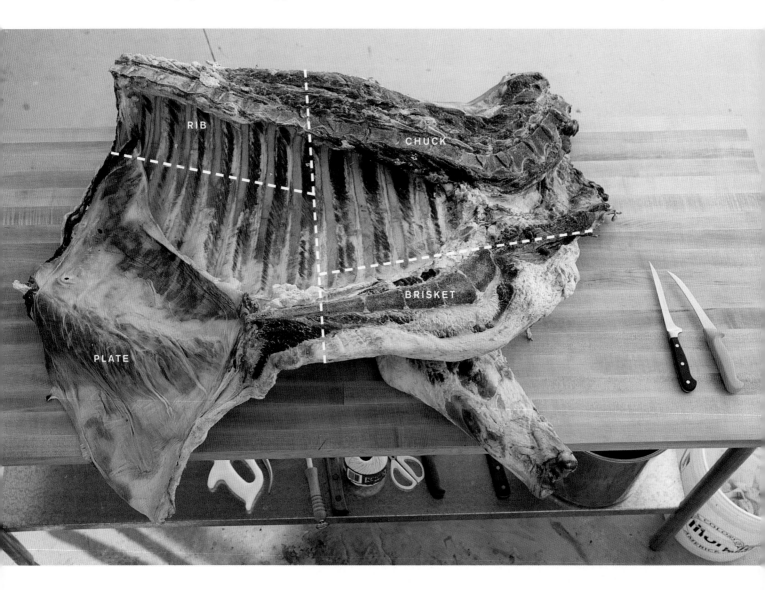

BREAKING DOWN A VEAL FOREQUARTER

The forequarter of veal differs from beef in that the brisket and plate (and optionally, the flank) are removed together as a breast. The following procedures focus on the industry standard methods, which result in a square-cut chuck, rib loin, and breast.

BREAST AND FORESHANK. The breast and foreshank are removed with a single cut through the arm and all the ribs, thus creating the first straight edge of the square cut.

CHUCK-RACK SPLIT. Separate the chuck from the rack by cutting between the fourth and fifth ribs. Use a breaking knife to make the initial cut between the ribs, starting at the bottom (ventral) edge of the ribs and heading toward the spine. Along the way you will run into the scapula. On some younger animals you may be able to cut through the scapula with your knife, but on most veal the ossification is too advanced and will require a saw.

FOREQUARTER

Remove the Outside and Inside Skirt

Prior to splitting the forequarter into primals, take a quick moment to remove the two skirt muscles. Position the forequarter so that the interior cavity is facing up, spine on the far side. The outside skirt should be easy to identify: it's the only muscle attached to the interior face of the ribs.

1. TRIM THE MEMBRANE.
The outside skirt, or diaphragm muscle, is enrobed in membranes. Remove the exterior membrane while the muscle is still attached. Trim the top edge of the muscle, just below where the two joining membranes end and the muscle begins.

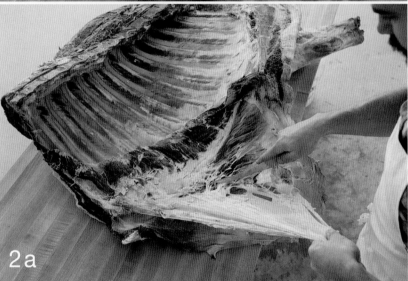

2. PEEL THE MEMBRANE.
Take the newly released edge of the membrane and peel it away from the muscle. Use your free hand to secure the muscle to avoid any tearing. Continue peeling the membrane away from the inside skirt.

3. SEPARATE THE OUTSIDE SKIRT.
Start at the anterior end and cut the outside skirt from the ribs. Keep your knife blade flat against the bones as you work.

4. SCORE THE INSIDE SKIRT.
Next, look to the bottom edge of the ribs for the plate portion of the inside skirt. Sever the edge with your knife, stopping before you reach the underlying membrane. Pull the muscle away and use your knife to cut any stubborn attachments.

FOREQUARTER

Split the Chuck from the Rib

The chuck and rib primals are split with a straight cut starting between the fifth and sixth ribs. Find the space by counting five ribs from the anterior end or six ribs from the posterior end. (Be aware that the first rib may be obscured, as it is small and often sits beneath fat and clotted blood deposits.)

1. SAW THROUGH THE SPINE.
Follow the space between the fifth and sixth rib to the spine and mark where you are going to saw. Saw through the spine, stopping when you reach muscle.

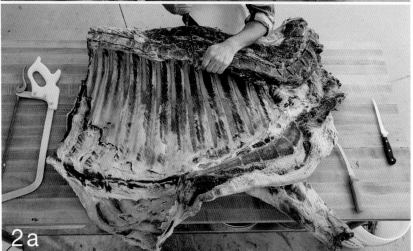

2. CONTINUE CUTTING WITH BREAKING KNIFE.
Sever the loin muscles and continue cutting between the ribs until you reach the sternum. If you are unable to cut through the scapula, then pull out the saw, make your way through the bone, and then return to the knife to finish cutting. Split the plate from the brisket by continuing the cut between the fifth and sixth ribs. Use your saw when you encounter bone.

3. SAW THROUGH THE STERNUM.

To finish the separation, saw through the sternum and cut any remaining connections between the two primals.

FOREQUARTER

Split the Rib and Plate

The desired length of your rib steaks determines where the rib and plate primals are separated. You determine this length based on the length of the rib bone (called a "tail"). If you are fabricating at home, I recommend that you leave about 3 inches, measured from the loin muscle, on both ends. (A rib primal destined for extended dry aging will need a couple extra inches to account for the trim loss.)

The rib loin has two ends: the chuck end (anterior) and the loin end (posterior). The chuck end is the larger of the two because it has more muscle groups, despite that the main loin muscle (*longissimus dorsi*) tapers in the direction of the chuck. Measure the decided tail length from the loin muscle. (For suggested measurements, and to find the right points, look for the fat indicators, shown in the accompanying photos.)

On the loin end, the point to cut is at the tip of a fat triangle that tapers toward the plate.

On the chuck end, the point to cut is in the middle of a fat deposit, located a few inches from the scapula.

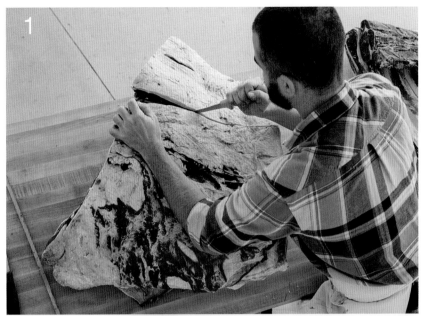

1. MAKE YOUR MARKS.
Mark your spots on both the chuck end and the loin end, and then connect the points atop the ribs to define your saw line, cutting down to the bone.

2. SAW THROUGH THE BONES.
Further processing of the rib and plate is on page 235.

SPLIT THE RIB AND PLATE

CREATING THE SQUARE-CUT CHUCK

THE SQUARE-CUT CHUCK is the industry standard for separating the chuck primal. This cut is done on a table or at a bandsaw. At this point, you have separated the chuck primal on the rib end by a straight cut between the fifth and sixth ribs. The next step is to separate the brisket and foreshank. Use a straight cut perpendicular to the face of the rib end to produce the characteristic "square-cut."

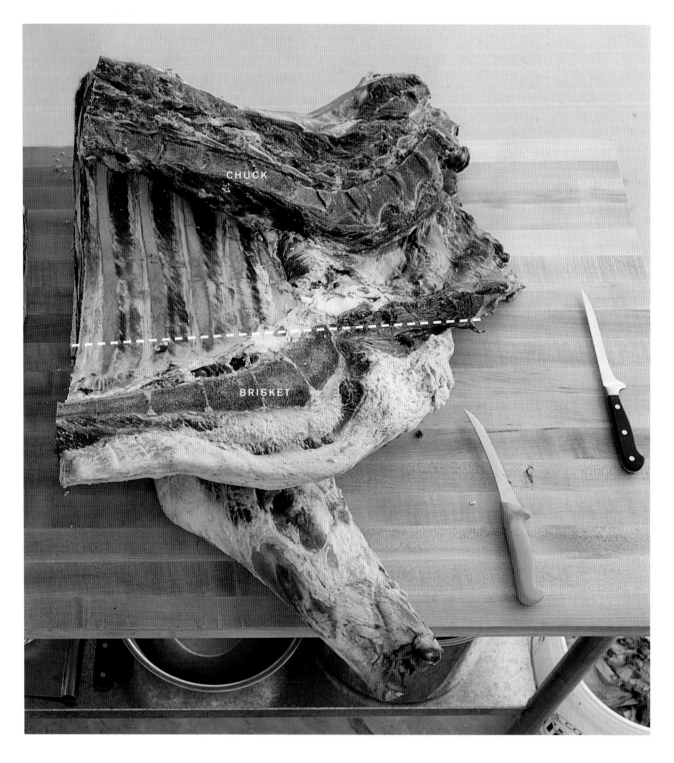

SQUARE-CUT CHUCK

Remove the Brisket and Foreshank

1. SCORE YOUR SAW LINE.
Place the chuck rib-side up. Align your cut with the bottom of the ribs, leaving the rib cartilage and sternum on the brisket.

2. SAW THROUGH THE RIBS.
Once you are through the bones, continue the separation using a long-bladed knife. At the anterior end you will run into the humerus. Saw through it, and then complete the removal with your knife.

SQUARE-CUT CHUCK

Remove the Brisket and Foreshank CONTINUED

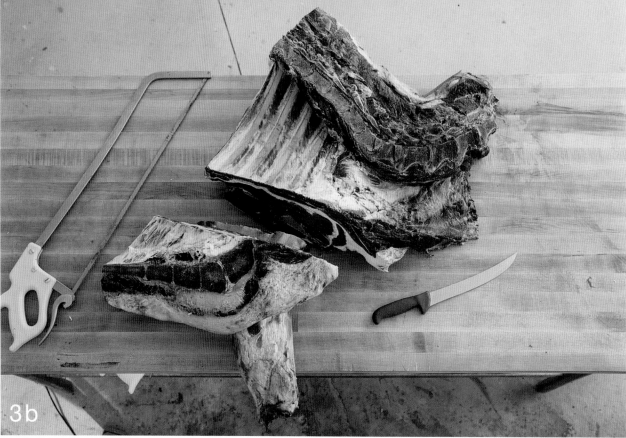

3. **SAW THROUGH THE ARM BONE.** Finish separation with your knife.

SQUARE-CUT CHUCK

Separate the Brisket and Foreshank

The brisket and foreshank are easily separated through a natural seam that is quite evident after they are separated from the square-cut chuck. If you have trouble finding this seam, apply pressure with your hands between the brisket and foreshank. The seam should be evident, starting at the "armpit." Follow the natural seam with your knife to separate the brisket and foreshank. Further processing information for both cuts starts on page 232.

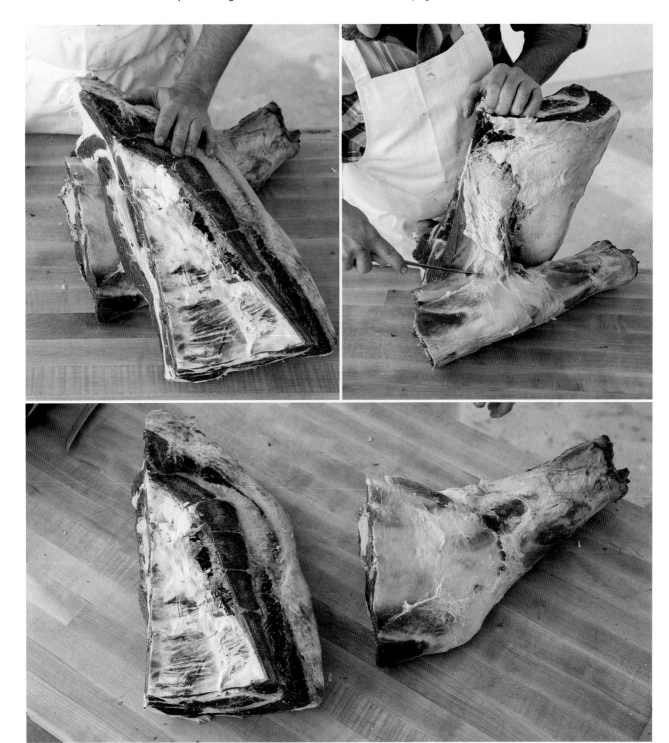

SQUARE-CUT CHUCK: THE QUICK BREAKDOWN

There are many ways to break down chuck. Although the two approaches detailed later in this chapter follow the natural seams of the muscle, there are also simpler methods. This simpler approach to breaking down a square-cut chuck gives you some of the best options for larger cuts, yet also provides adequate amounts of grind and stew meat. This approach is ideal for those who do not want to focus on individual muscle groups or advanced seam butchery.

- Make sure to trim away excess connective tissue from meat designated for grind, and remove any excessive fat deposits for stew meat. In addition, cut out and discard any remaining meat that looks unpalatable.

- Start with the square-cut chuck, ribs facing down. Separate the arm and block from the rest of the chuck by making a cut that is parallel to the spine about one-third up from the bottom edge, around where the scapula and humerus join. Cut through muscle, saw through bone, until you reach a seam above the chuck rib meat. Separate the arm by following the seam atop the rib meat. Use the arm as a bone-in pot roast, or bone it out for a rolled pot roast or grind.

- Bone out the whole rib section, following the flat of the ribs with a boning knife. Save whole for an incredible roast, or cube for the best stew meat on the animal.

- After handling the chuck ribs, cut a few chuck steaks off the rib end of the chuck. Make them about 1 to 1½ inches thick. (The first three steaks can be considered for Delmonicos when boned.) The diversity of muscle groups in these steaks makes them a delicious alternative to the ribeye and strip steaks, which are dominated by the main loin muscle, the *longissimus dorsi*.

- Continue by cutting off 7-bone steaks from the rib end, 1½ to 2 inches thick. Stop when you are close to the joint of the scapula. Bone the rest and use for stew, grind, or roll sections for pot roasts. The neck bones are great for use in stock and soup.

Square-Cut Chuck (Seamed Breakdown)

Each of the muscles in the chuck have individual characteristics. These characteristics are revealed only when the muscles are removed one at a time. This approach of butchery depends on the natural seams between these muscles.

SQUARE-CUT CHUCK

Remove the Humerus

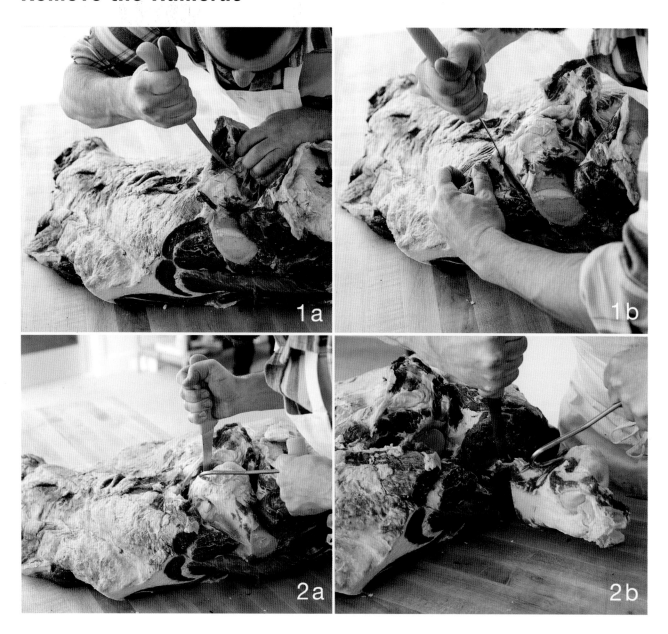

1. FIND THE SEAM.

Removing the humerus gives you easy access to the shoulder clod subprimal. Start with the chuck bone-side down and approach the humerus through a natural seam between the clod heart (the beef triceps) and surrounding muscles. Start where the bone was exposed after the brisket separation. Separate both sides of the bone until you reach the juncture with the scapula.

2. EXPOSE AND REMOVE THE HUMERUS.

Sever the tendons that keep the humerus attached to the scapula. They are thick and will require a sharp knife and some pressure. Do not try to force your way through bone; make sure you are working between the bones and through the connective tissue. Work your way down the underside of the bone until it is fully released.

SQUARE-CUT CHUCK

Separate the Shoulder Clod and the Chuck Tender

The shoulder clod is one of the two chuck subprimals and is composed of mainly three muscles: the top blade, shoulder tender, and clod heart. The best way to approach the removal of the shoulder clod is from the exterior. Stand at the spine (dorsal edge); this gives you easy access to the top blade, a muscle that can be tricky to work with due to the shape of the scapula and scapular spine. On the opposite side of the scapular spine is the chuck tender.

1. **START SEPARATING THE CHUCK TENDER.** Find the scapular spine by feeling for the ridge of bone that sits just underneath the outer layer of the carcass. This ridge is demarcated by a white line between the two exterior scapular muscles. Use a boning knife and follow the ridge, initially on the front (anterior) edge to begin releasing the chuck tender. Make sure to score the bone sheath as well. Do not worry if the chuck tender is slightly damaged when you trace the scapular spine. Working on the chuck tender side first is advantageous because it ensures that you find the proper edge of the scapular spine while you work next to a low-value muscle. This method reduces the risk of damaging more valuable muscles, such as those in the shoulder clod.

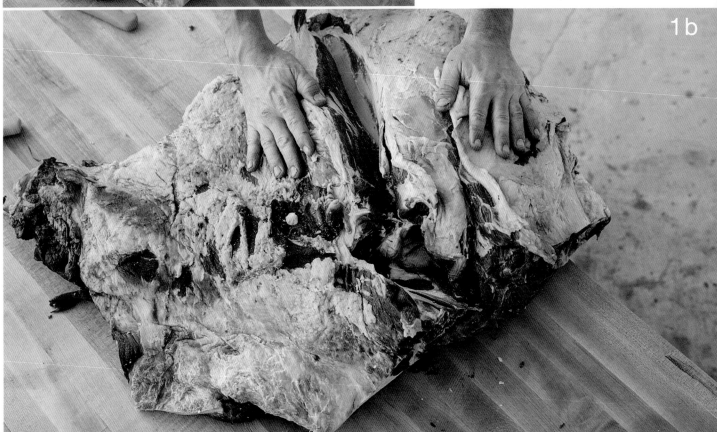

SQUARE-CUT CHUCK

Separate the Shoulder Clod and the Chuck Tender CONTINUED

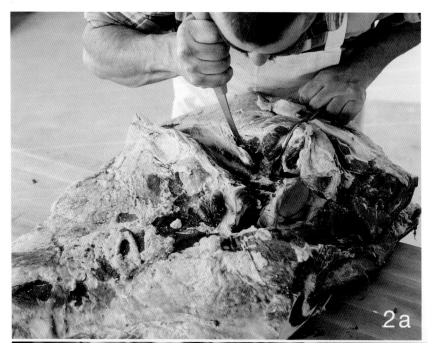

2. START SEPARATING THE TOP BLADE.
The scapular spine slopes on its rear (posterior) edge. When tracing the spine on that side, angle the tip of your knife slightly toward the spine. Start near the joint end of the scapula and cut along the anterior edge of the scapular spine. Make sure to cut through the underlying bone sheath. This cut will allow for the peeling of the muscle group later.

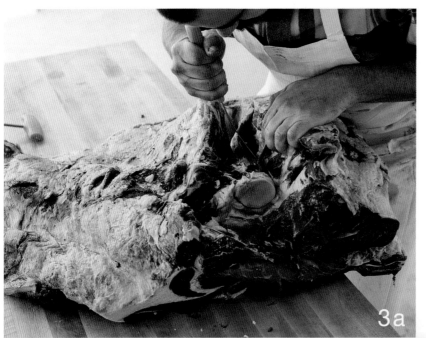

3. PEEL THE TOP BLADE.

Sever the connective tissue where the top blade meets the joint of the scapula, staying close to the bone. With the bone sheath properly scored along the scapular spine, you should be able to peel the top blade away from the spine toward you.

SQUARE-CUT CHUCK

Separate the Shoulder Clod and the Chuck Tender CONTINUED

4. **SEPARATE THE WHOLE SHOULDER CLOD.**
Follow the underside seam of the clod heart and shoulder tender. Pull the latter, which sits next to the under blade flap, to expose the seams. To finish removal, follow the seams all the way to the end of the scapula.

5. **PEEL THE CHUCK TENDER.**
Cut across the thicker head of the chuck tender near where it meets the scapula. Using a boning hook, peel the chuck tender away from the scapula, and sever any stubborn attachments as you go.

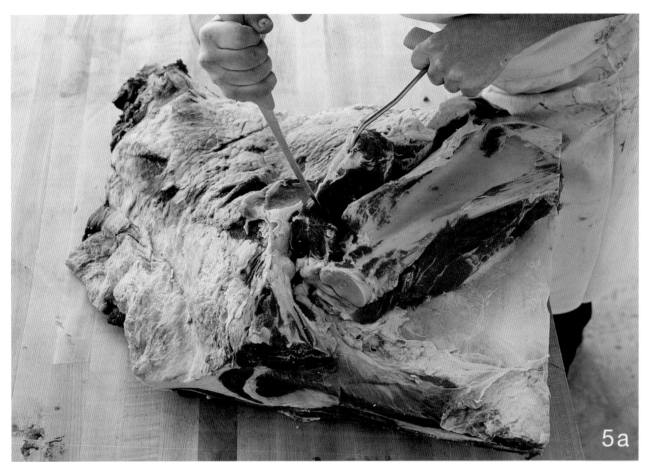

5a

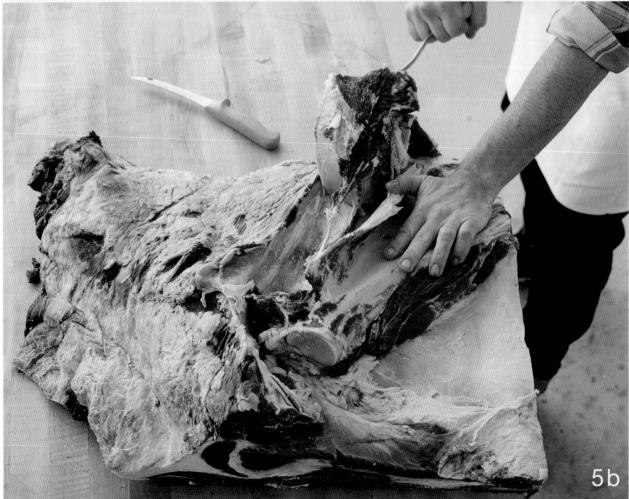

5b

SQUARE-CUT CHUCK

Seam Out the Shoulder Clod

The shoulder clod can continue to be broken down into its constituent parts, all by following natural seams. Use small exploratory cuts at first to ensure that you are following the natural seams.

1. BEGIN BY REMOVING THE SHOULDER TENDER.

Find, and follow, the seam between the shoulder tender and the underlying thick silverskin of the clod heart. Finish separating the shoulder tender from the small side muscle. Leave the small muscle attached to the clod heart, or remove it for grind.

2. SEPARATE THE TOP BLADE.

The top blade is connected to the clod heart through a single seam of thick connective tissue. Find that seam with your knife to separate the two.

SQUARE-CUT CHUCK

Remove the Scapula

At this point in the process, the scapula is still attached through a thick fascial layer that runs along the underside and around both edges. The muscle that lies underneath is the under blade flap.

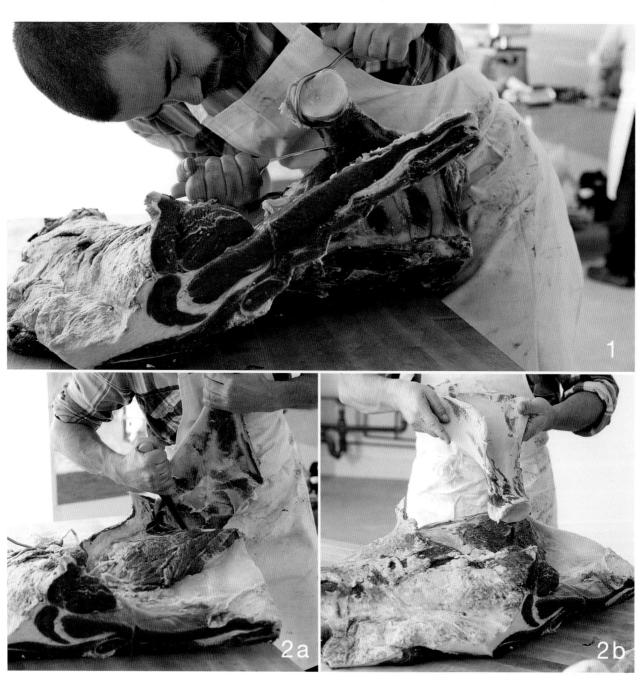

1. SCORE THE CONNECTIVE TISSUE.

Start at the joint of the scapula and, while pulling up on the bone, sever the underside connective tissue that attaches to the muscle. Then score both underside edges of the scapula using a boning hook.

2. PEEL AWAY THE MUSCLE.

Remove the under blade flap by following the flat of the bone with your knife, either getting underneath the bone sheath with the back of the blade or cutting across using the blade itself.

BEEF BUTCHERING

Boning and Seaming the Chuck Roll

At this point in the process, you have removed the humerus and scapula bones as well as the shoulder clod subprimal and lower arm muscles. The remaining bones in the chuck are the neck (cervical) vertebrae and the five thoracic vertebrae with their corresponding ribs. The remaining muscles are the chuck roll subprimal, neck, some remnants of the brisket, and other small arm muscles. We start by boning the neck, making the following step — boning the chuck roll — much easier. We finish the process by seaming the constituent parts of the chuck roll for various preparations.

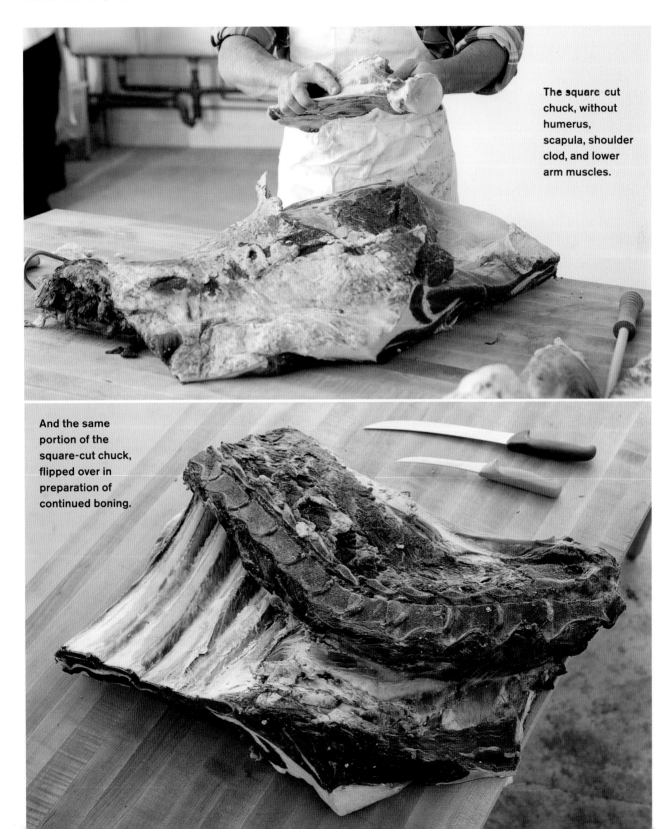

The square cut chuck, without humerus, scapula, shoulder clod, and lower arm muscles.

And the same portion of the square-cut chuck, flipped over in preparation of continued boning.

SQUARE-CUT CHUCK

Boning the Neck

Boning the neck may be one of the most difficult butchering tasks, as the vertebrae are intricate and vary in shape from one to another. Study the skeleton on page 138 to better understand the shapes that you will encounter. Save the neck bones for stock or soup. Alternatively, after separating the last neck bone you can cut straight through the neck meat for a bone-in neck, which then allows for other preparations (see page 224).

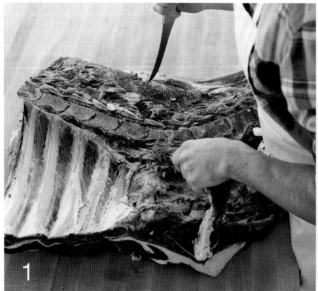

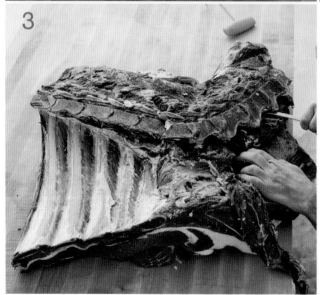

1. UNLINK THE CHAIN.

Flip over the chuck, bones facing up. There is one neck muscle, sometimes called the chain, which runs on the interior edge of the rib (thoracic) vertebrae and continues into the neck. Start at the posterior end and, following the shape of the bones, separate the muscle from the rib and neck vertebrae.

2. SEPARATE THE CERVICAL AND THORACIC BONES.

Find the last neck (cervical) vertebra by counting seven bones from the tip of the neck. Work your knife between the last neck bone and first rib vertebra, following their shapes by cutting where there is less resistance.

3. RELEASE THE UNDERSIDE FIRST.

Do all the neck boning from the last neck bone, working your way toward the front. Your goal is to release both sides of the neck bones and then work your way down the middle, pulling on the last bone to apply outward pressure to help the separation. Begin by releasing the muscles on the underside (ventral) portion of the bones, and then follow with the top (dorsal) side. This process, as with boning the entire neck, will be largely exploratory the first several times you try it. The work will become easier after you learn the shape of the bones.

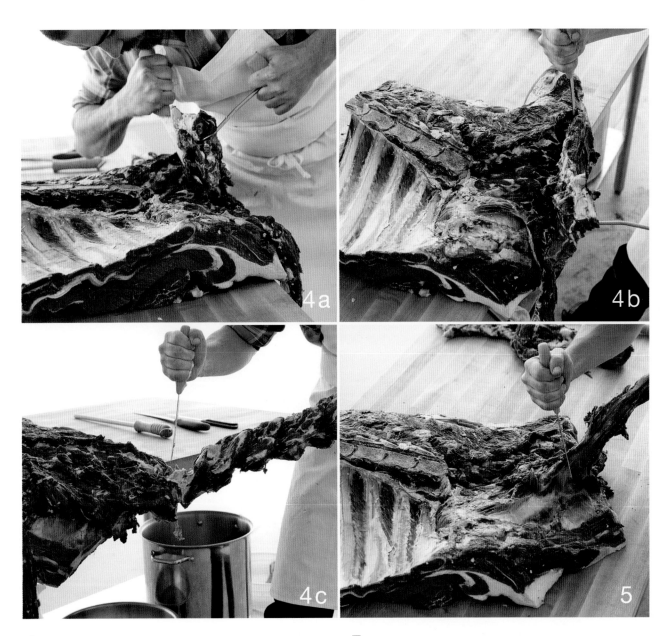

4. WORK DOWN THE MIDDLE.

Pull up on the last neck bone and then follow the underside of the bones, which connect the two sides that you previously separated. Continue down the length of the bones, following their shape and working from side to side, until they are all released.

5. FINISH REMOVING THE CHAIN.

Work your way down the seam of the chain muscle until it's separated, and then clean the muscle for grind.

SQUARE-CUT CHUCK

Boning the Chuck Roll

Now that the neck bones are removed, you have a bone-in chuck roll. Before boning you have the option of cutting off the second through fifth ribs for chuck short ribs. (The main muscle of the ribs, also known as the under blade roast, is delicious either on the bone or boneless.) To separate the short ribs, saw perpendicular to the ribs a couple of inches from where they join with the spine. Stop when you reach muscle, and then finish the cut with your knife.

The chuck, after removal of the neck bones.

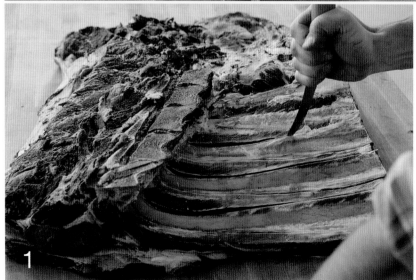

1. TRACE THE RIBS.

Begin boning by tracing the edges of the ribs with the tip of your knife, separating the intercostal rib meat (the thin muscles between the bones) from the bones.

2. SEPARATE THE UNDERSIDE.

With the edges of the intercostal meat separated, lift up the bottom edge of the fifth rib and follow the flat underside toward the vertebrae. Repeat with each rib, one by one.

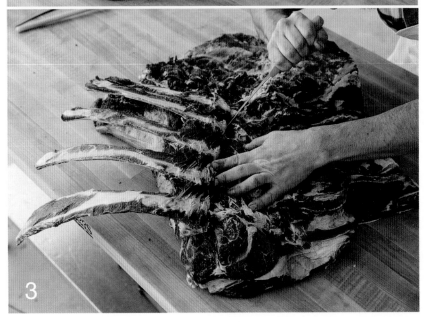

3. REMOVE THE VERTEBRAE.

Make sure the ribs are separated all the way down to the juncture with the vertebrae. Note how the vertebrae jut out at the base of the ribs. Work the full length of the vertebrae. Cut out and around the outcropping, staying close to the bone, until you hit the flat of the feather bones. To finish boning the chuck roll, push the bones away from you while you follow the flat of the feather bones (the flat protruding portion of the vertebrae that is the result of splitting the carcass in two) with your knife.

SQUARE-CUT CHUCK

Boning the Chuck Roll CONTINUED

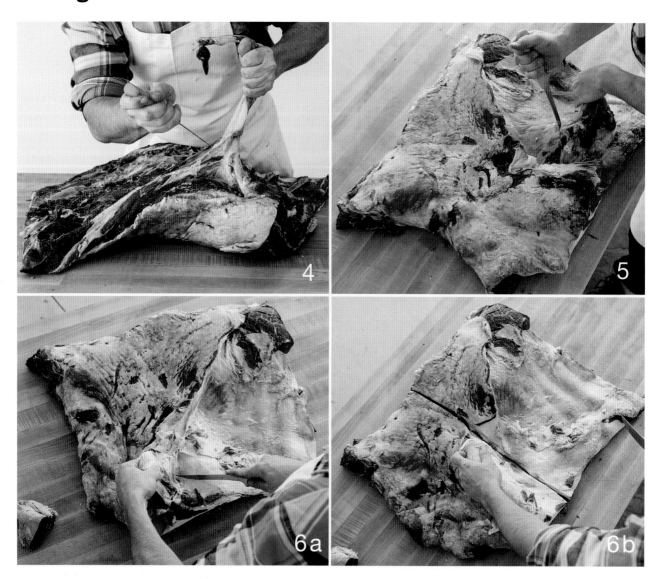

4. REMOVE THE NUCHAL LIGAMENT.
The thick nuchal ligament, composed mostly of elastin, runs along the dorsal edge of the neck and chuck roll. Remove it by pulling it away from the muscles and severing its connections.

5. SEPARATE THE UNDER BLADE FLAP.
The under blade flap, which sits just underneath the scapula, is easily removed now that the bone is gone. Flip the chuck over and remove the steak through the underlying seam.

6. SPLIT THE CHUCK ROLL AND NECK.
At the base (ventral edge) of the chuck roll are some remnants of the brisket. Follow the seam beneath those brisket muscles so that you can fold them over and keep them with the neck end of the chuck when you separate the two sides. Split the boneless chuck roll from the boneless neck, cutting at the point where the neck bones ended. Stay parallel to the rib (posterior) end of the chuck.

SQUARE-CUT CHUCK

Seaming the Chuck Roll

There are three notable cuts from this boneless subprimal: chuck eye roll, under blade roast, and the under blade flap. I prefer to separate them for their individual qualities, of which there are many. The chuck roll has the most muscles of any subprimal, so don't be put off by the initial uncertainty in finding the right seams. Further processing for each of these cuts is available starting on page 216.

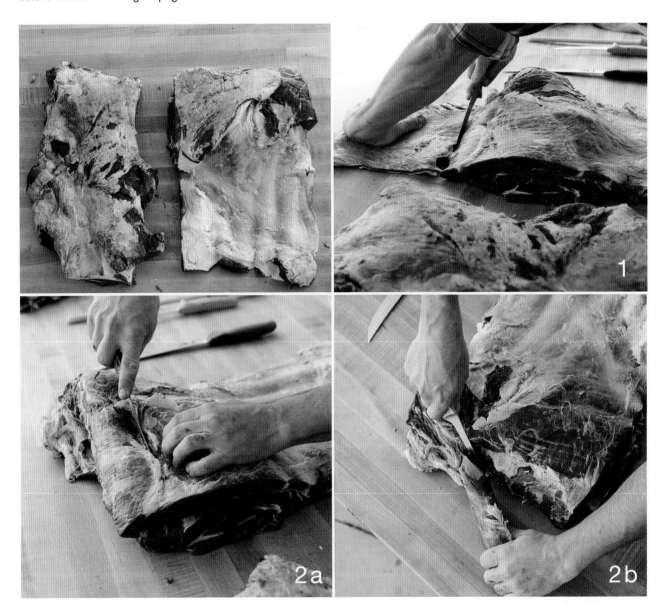

1. REMOVE THE *TRAPEZIUS*.
Lying atop the chuck roll is the exterior fat and a thin muscle known as the *trapezius*. Peel away this thin layer using the underlying seam.

2. REMOVE THE *RHOMBOIDEUS*.
Although the *rhomboideus* is technically part of the under blade roast muscle group, I recommend that you remove it at this point. Find the seam of this tapering muscle and follow it for removal. Save it for stew or the grind pile.

SQUARE-CUT CHUCK

Seaming the Chuck Roll CONTINUED

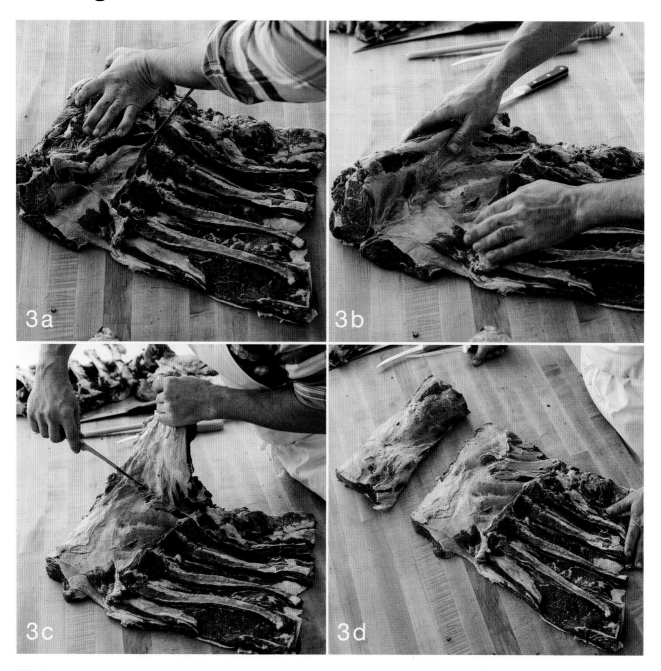

3. SEPARATE THE CHUCK EYE ROLL.

The chuck eye roll is a grouping of four muscles. Together, this grouping forms one of the most flavorful roasts in the entire carcass. The chuck eye roll can also be seamed further for smaller roasts or steaks. The seam sites are located on the underside of the chuck roll, near where the rib–vertebrae connection used to reside. Flip the chuck roll over and locate that seam. Follow the seam over the underlying muscles (the under blade roast), erring on the side of the roast rather than on the more valuable chuck eye roll.

4. SEPARATE THE UNDER BLADE ROAST.

The intercostal meat and some of the thinner muscles of the ribs still sit next to the remaining two muscles of the under blade roast: the *splenius* (on top) and the *serratus ventralis*. Find the seam, which follows the tapering bottom edge of *serratus ventralis*, to remove the thin rib muscles.

SQUARE-CUT CHUCK

Seaming the Chuck Eye Roll

There are four main muscles in the chuck eye roll: *complexus*, *multifidus dorsi*, *spinalis dorsi*, and *longissimus dorsi*. The first three are considered some of the top muscles in the entire carcass when ranked for tenderness; each of them can be separated through natural seams. The *longissimus dorsi* is the same muscle that makes up the eye of the rib and short loins; in the chuck it gets smaller and tougher as it heads toward the neck. These qualities make it less desirable than the cuts within the rib and short loin primals.

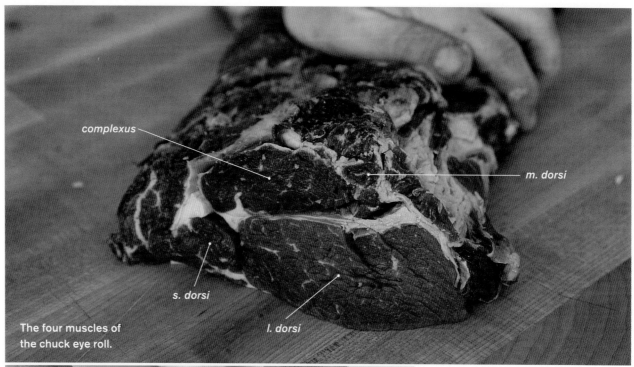

The four muscles of the chuck eye roll.

1. SEPARATE THE COMPLEXUS.

Place the chuck eye roll so that the medial face (the face that was against the feather bones) is facing up. Begin by finding the seam that separates the *complexus* muscle from the underlying *multifidus dorsi* and *longissimus dorsi* (which are grouped together). Follow that seam to separate the ventral side of the *complexus*.

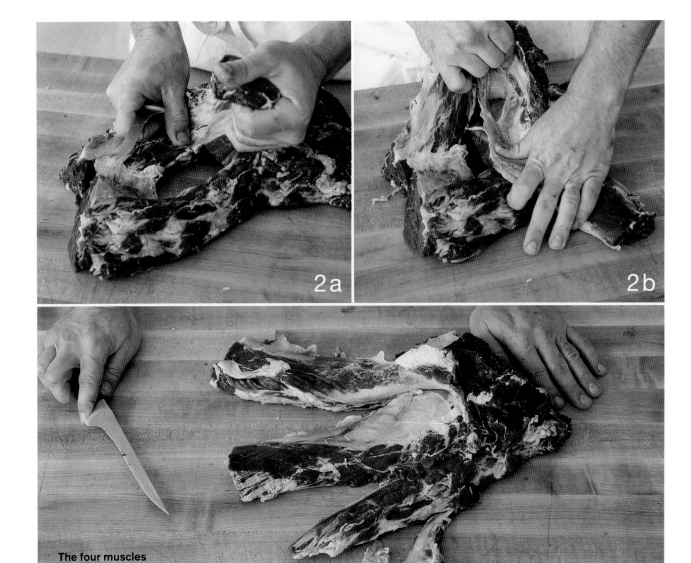

The four muscles (from top to bottom): *spinalis dorsi, complexus, multifidus dorsi,* and *longissimus dorsi.*

2. SEPARATE THE *SPINALIS DORSI*.

On the other side of the *complexus* is the seam that joins it to the *spinalis dorsi* muscle. It's a seam that can be pulled apart mostly by hand, but if you do so make sure to pin one of the muscles against the table to prevent tearing.

3. SEPARATE THE LAST TWO MUSCLES.

The remaining two joined muscles are the *multifidus dorsi* and the *longissimus dorsi*. The *multifidus dorsi* is a thin muscle that formerly sat against the spine. Distinguishing it from the loin muscle should be relatively easy to do. Find the seam between the two and use it to separate them.

BREAKING DOWN THE HINDQUARTER

THE HINDQUARTER IS made up of the flank, round, and loin. There are a few different approaches to breaking down this quarter into primals and subprimals. One method may appeal to you more than another, so feel free to pick and choose based on your preferences. However, be sure to read through this whole section before you butcher, as variations are mentioned throughout.

HINDQUARTER

Remove the Kidney Fat

Remove the kidney (if it's still attached) and kidney fat, saving the fat for grind. You may need to use your knife to release some of the hardened fat, but do so carefully, as the valuable tenderloin lies beneath. Look for the hanger steak — it will be attached to one of the carcass sides, not both — and cut it away, saving it for further processing. (For those planning on aging the loin, leaving the kidney fat on will help protect the tenderloin from drying out. In that case, just remove the kidney itself.)

HINDQUARTER

Remove the Flank

Removing the flank is an easy first step that reduces the bulk and weight of the hindquarter, which is especially helpful if you are working on a table. There are three cuts that can be pulled from the flank: sirloin flap, inside skirt, and flank steak. Each cut has unique characteristics, and while many processors will automatically toss the first two into their hamburger grind, I strongly encourage you to prepare them and judge for yourself.

The flank starts near the sirloin tip and continues all the way to the thirteenth rib. Removal will work in the same direction. The ideal approach is threefold: follow natural seams and muscle curvature along the round and sirloin, find the edge of the sirloin flap, and then cut parallel to the vertebrae while keeping your distance from the main loin muscles.

1. CUT ALONG THE ROUND AND SIRLOIN.

Hold the flank flap near the round and feel for the juncture between the thin abdomen and large leg muscles. Cut along that juncture, where the fat visibly ends and the round meat begins.

2. FIND THE SIRLOIN FLAP.

Pull down on the flap (a boning hook works well for this) while cutting through the fat, following the curvature of the round muscles, until you reach the sirloin flap, a muscle that has an apparent graininess compared to the firm muscles of the round and sirloin.

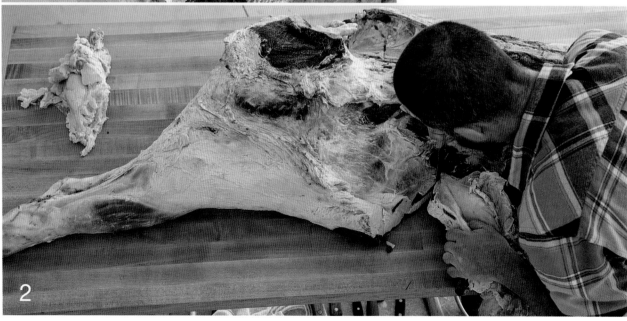

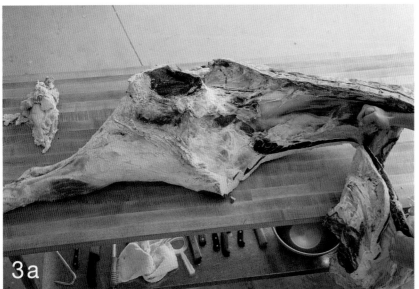

3. CUT PARALLEL TO THE VERTEBRAE.

When you reach the sirloin flap, turn your knife and cut toward the rib, roughly parallel with the vertebrae. Keep the knife on the side of the sirloin flap muscle, leaving some behind if need be, to avoid cutting too deep and into the loin. Saw through the last rib to fully separate the flank.

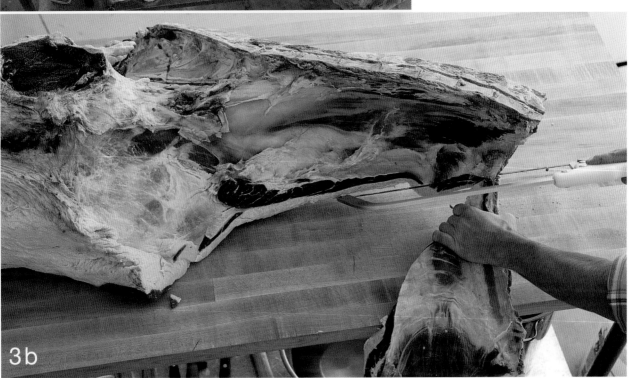

PROTECTING THE LOIN

While not shown in the photos here, when you are learning the contours of the muscles surrounding the flank, especially the valuable loin, it is helpful to initially mark the line for cutting that will run parallel to the loin and vertebrae. Look at the rib (anterior) end and choose the distance between the loin muscle and where you want to cut, also known as the tail length. A common distance is around 2 to 4 inches, but feel free to choose based on preference. Remember that the loin muscle tapers as it gets closer to the neck, so the end that you can see (the anterior end) is going to be smaller than the end near the sirloin (the posterior end). Measure the distance from the loin muscle, mark it with your knife, and saw through the thirteenth rib at that point, which is called your rib mark.

HINDQUARTER

Separate the Three Main Flank Cuts

Now that you have the flank removed, you can quickly separate the three main cuts. All of the flank muscles are large-grained and highly susceptible to tearing if pulled on too hard. For this reason, when you encounter considerable resistance use your knife to sever connections. Save all the muscles as is until you want to use them, at which point you can clean and process them for cooking. (Further processing instructions begin on page 281.) This keeps the valuable (and edible) muscle protected by the membranes and trim, even when freezing.

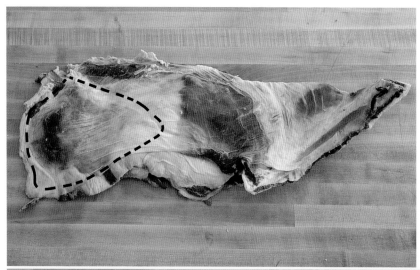

1. FIND THE FLANK STEAK.
Identify it and trace around it with the tip of your knife, staying shallow.

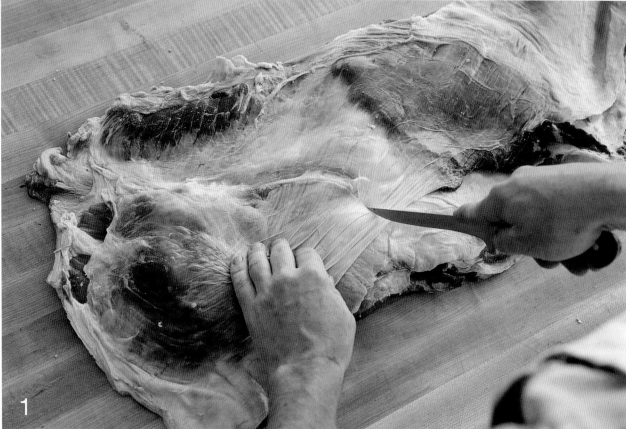

BEEF BUTCHERING

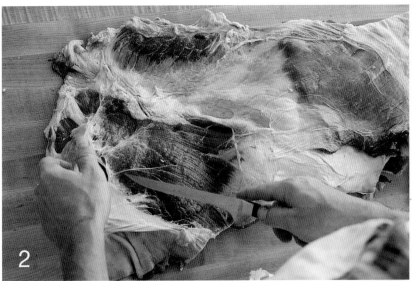

2. PEEL AWAY THE OUTER MEMBRANE.

Pull the membrane away from the muscle, starting at the edges you just cut through. Use your free hand to pin the muscle down to the table to avoid tearing the muscle. Stubborn connections can be severed with the knife. Cut away the membrane after its removal.

3. PEEL THE MUSCLE FROM THE UNDERLYING MEMBRANE.

Using your fingers, get between the pointy end of the muscle and the underlying membrane, and then peel it away from the fatty membrane beneath it. This step can be mostly done by hand, though you will run into areas that are more easily released with a knife. Finish by cutting across the bottom where the muscle tapers into fat.

SEPARATE THE THREE MAIN FLANK CUTS

HINDQUARTER

Separate the Three Main Flank Cuts CONTINUED

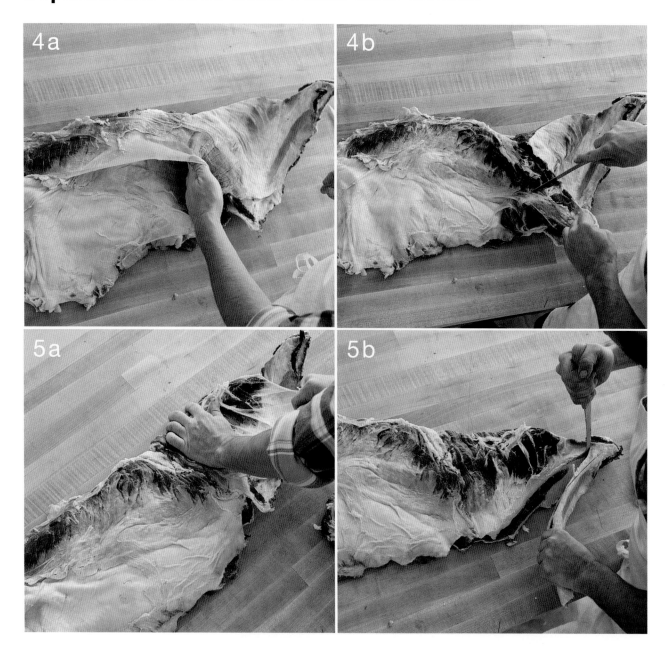

4. REMOVE THE INSIDE SKIRT.

The inside skirt spans two primals, the plate and the flank, and is not to be confused with the outside skirt, also known as the diaphragm. Removing the inside skirt, like all flank muscles, will require little knife work; it should peel away from its connections with little resistance. Identify it and pull it away from the underlying membrane. Be careful not to tear the muscle, and use your knife when appropriate.

5. PEEL THE TOP OF THE SIRLOIN FLAP.

The sirloin flap is the largest of all the flank muscles. It sits beneath part of the inside skirt, hence it is removed last. Follow the same method you used for the other flank muscles, identifying the full muscle and then making shallow cuts to sever attachments while pulling to release the muscle from the underlying membranes. Start by peeling the top membrane away, near the last rib. Remove the rib if you haven't done so already.

BEEF BUTCHERING

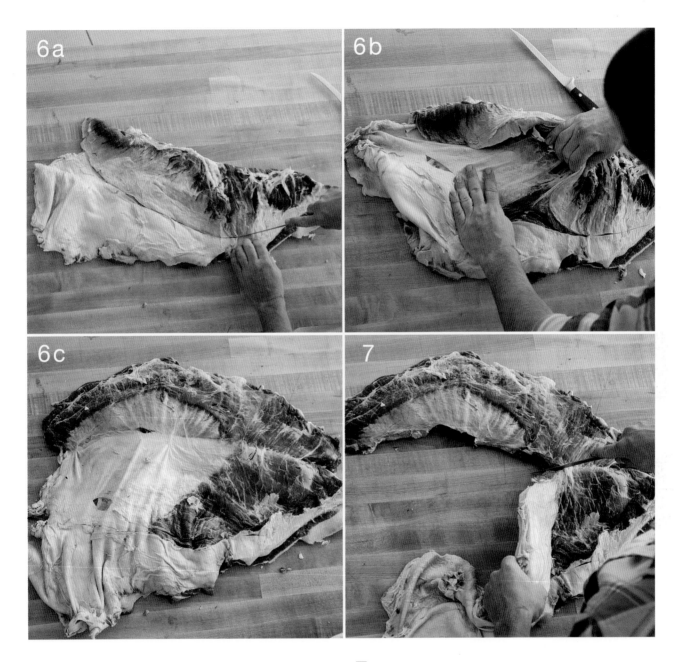

6. OUTLINE AND RELEASE THE UNDERSIDE.

Trace the bottom edge of the sirloin flap with your knife tip. Get your fingers beneath the edge and peel the muscle away from the underlying, thicker membrane. Underneath the sirloin flap, near where the rib resided, is another small flank muscle that should stay attached to the flank.

7. FINISH REMOVAL.

To finish the separation, cut along the edge where the thickest part of the sirloin flap attaches to the sirloin.

HINDQUARTER

Separate the Three Main Flank Cuts CONTINUED

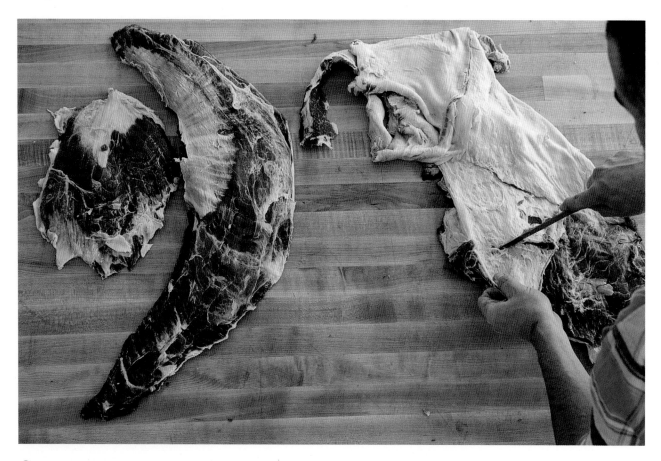

8. TRIM THE FLANK.
The remaining portion of the flank should be trimmed of any glands and other inedibles, including too-aged or dried meat. Save this portion of flank for grind. The exterior layer of meat – its proper name is *cutaneous trunci*, but it has many monikers, including wiggle meat, fly shaker, and elephant ear – should be removed and discarded. The fat in the flank, sometimes referred to as cod fat, is an excellent addition to lean ground beef to help increase the fat ratio.

HINDQUARTER

Split the Loin from the Round

Saw through the pelvis and sacral vertebrae to separate the round from the loin. This leaves the sirloin on the short loin, making it the full loin. A cut like this is especially common when working solely on a table, as it leaves two primals that are both manageable to lift and manipulate.

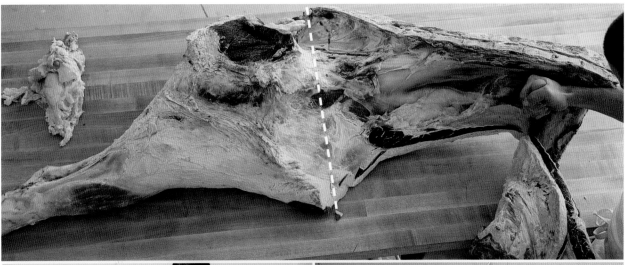

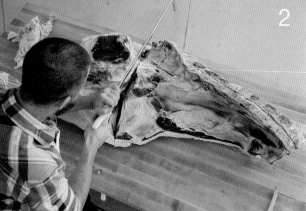

1. IDENTIFY THE RIGHT SPOT TO SAW.

Find the first sacral vertebra, which is the first vertebra that angles toward the tail after the last lumbar vertebra. Count out four or five vertebrae heading toward the tail, and mark that point with your knife. Now find the edge of the aitchbone; mark a point two or three finger-widths away from its edge and connect it to the point on the vertebrae, scoring the line with your knife. This is your saw line, and it should be relatively perpendicular to the loin and lumbar vertebrae — this cut will prevent waste when cross-cutting large sirloin steaks or roasts.

2. SAW AND CUT THROUGH.

Saw through the bones and, once you reach flesh, continue by using a breaking knife or other long-bladed knife. Make the knife cut with a smooth motion, rather than sawing through, to ensure that the face of your round and sirloin are clean.

Breaking Down the Loin

The loin is the most valuable primal in the entire animal, due mostly to the existence of the tenderloin. The breaking down of the loin presents one of the major either/or decisions that you will make when butchering: remove the tenderloin or cut T-bone and porterhouse steaks.

For those wanting to cut T-bone and porterhouse steaks, skip to page 197 and continue with separating the sirloin. Then continue on to cutting T-bone and porterhouse steaks following the instructions on page 262. Otherwise, continue on for full tenderloin removal.

LOIN

Remove the Tenderloin (Optional)

Removal of the full tenderloin will give you the option of cutting roasts or filet mignon, and this operation is performed prior to splitting the short loin and sirloin. By removing the tenderloin you eliminate the option of cutting porterhouse or T-bone steaks, because the tenderloin is part of those cuts. Instead, a short loin without the tenderloin will provide strip steaks or loin roasts.

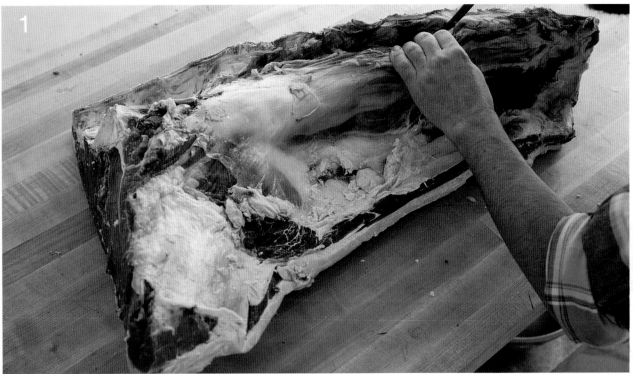

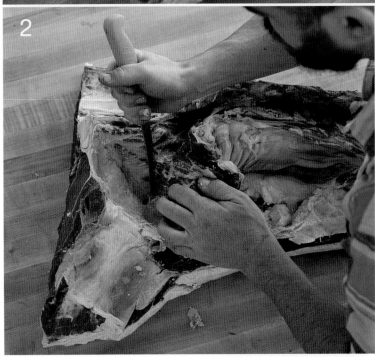

1. CUT ALONG THE VERTEBRAE.

Remove any remaining kidney fat that surrounds the tenderloin. Notice where the tenderloin connects to the last lumbar vertebra. Beginning there, use a boning knife to follow the contour of the vertebrae while staying close to the bone. Work your way toward the tail and begin to release the muscle from the bones.

2. RELEASE THE BUTT.

The tenderloin butt sits in a depression within part of the pelvis. To peel out of the butt, start by cutting along the edge of the pelvis that sits next to the tenderloin, following the contour of the bone and pulling the muscle away as you progress. Stay shallow to avoid cutting into the underlying sirloin.

LOIN

Remove the Tenderloin (Optional) CONTINUED

3. PEEL THE TENDERLOIN BACK.
Once the butt is released, use your hand to peel it toward you and out of the pelvis. Notice the small muscle that juts out from the butt of the tenderloin, commonly referred to as the foot. Leave it attached to the butt and use your knife to coax it out of the depression where it sits, and then cut it free from any connections.

4. CUT MEAT AWAY FROM THE FINGER BONES.
Once the entirety of the butt end is free, you can continue to peel it away from the finger bones. If you run into stubborn areas, do not pull harder; doing so risks damage to the tenderloin. Instead, switch to the knife and use it to scoop the muscle away from the finger bones of the lumbar vertebrae. Be careful not to cut between the bones and damage the adjacent loin muscle. Additional processing of the full tenderloin is covered on page 272.

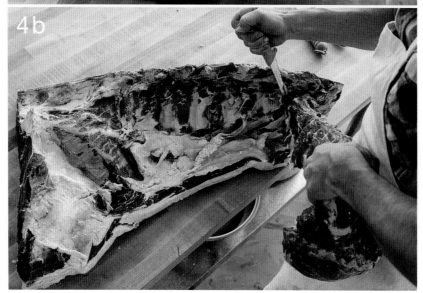

LOIN

Short Loin

The short loin is separated from the sirloin with a cut that is made adjacent to the front edge of the pelvis — essentially the top of the hip bone. Fortunately, this point will roughly align with the middle of the sixth, or last, lumbar vertebra, giving you an easy starting point. If you are separating without a saw, use the space between the fifth and sixth lumbar vertebrae. While the loin shown here has had its tenderloin removed, these instructions also apply for those looking to cut T-bone and porterhouse steaks.

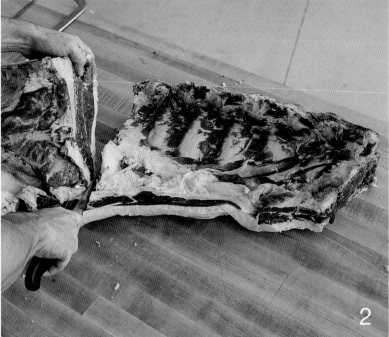

1. FIND THE MIDDLE OF THE SIXTH LUMBAR VERTEBRA.
Saw through the vertebra until you reach muscle, then stop.

2. CUT TO FINISH THE SEPARATION.
Use a breaking knife to cut from the loin out through the flank end. Stay perpendicular to the spine. If you hit bone before the flank, it means you have reached the edge of the pelvis. Knife around it rather than trying to cut through it.

LOIN

Boning the Sirloin

The sirloin subprimal comes from a weight-bearing area of the animal — meaning the muscles were hard at work, giving the meat a fabulously deep flavor and yet maintaining an overall tenderness. This combination provides great palatability. The sirloin is separated into two muscle groups: the top sirloin butt and the bottom sirloin butt. Here we will separate the muscles that make up the bottom sirloin butt, leaving us with the whole top sirloin butt (see page 202). The instructions that follow assume that you have removed the tenderloin butt from the sirloin. If not, follow the instructions for removing the butt end of the full tenderloin (see page 195) before continuing.

The sirloin is an interesting cut to bone due to the shifting shape of the ilium bone and the connection to the sacrum. The sirloin's main attachment is to the concave surface of the ilium. As you separate the two, stay close to the bone while trying to pull it away from the muscle; otherwise you may end up cutting into the top sirloin.

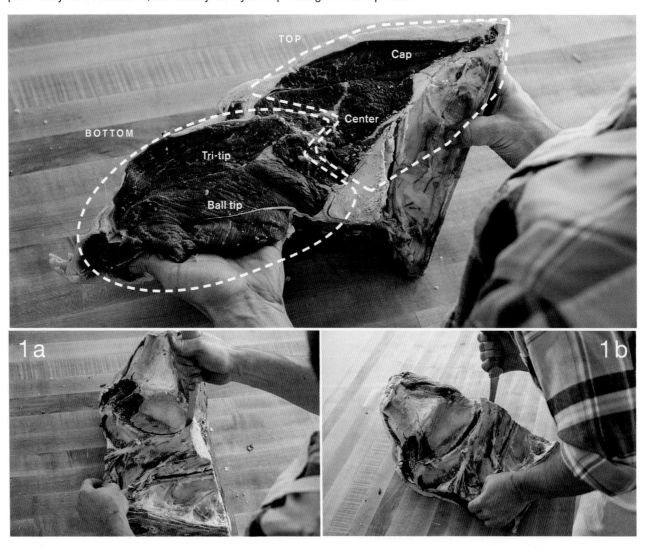

1. TRACE THE AITCHBONE.
Cut along the top edge of the ilium, starting at the loin (anterior) end. Then cut along the membrane that connects the aitchbone and sacrum.

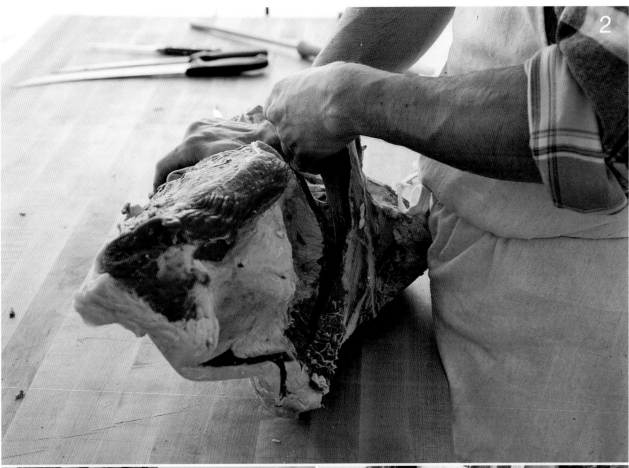

2. FOLLOW THE CONCAVE SURFACE.

Continue to follow the ilium bone with your blade, using its flexibility to your advantage. Keep your blade tip close to the bone as it slopes. You should get to a point where you can begin to peel the bone away from the muscle.

3. SEPARATE THE SACRUM.

With the ilium mostly separated, cut along the connection to the sacrum, first the inside and then through the fat cap on the outside.

BONING THE SIRLOIN 199

LOIN

Boning the Sirloin CONTINUED

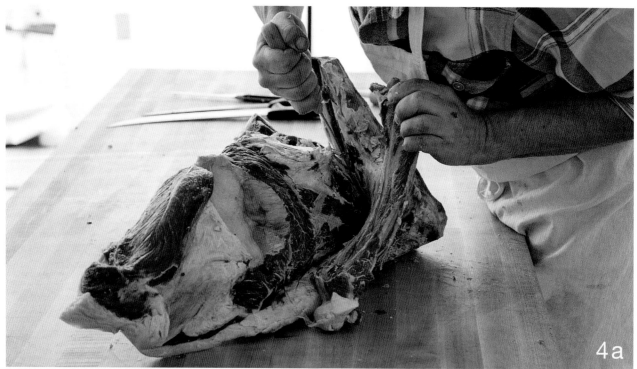

4. **FINISH PEELING AND REMOVAL.**
Continue following the slope of the ilium, pulling the bone away from the muscle. Sever any of the remaining connections along the loin (anterior) end. There is a cavity between the ilium and the sacrum that contains a small portion of meat (the remainder of the *longissimus dorsi*). You can't remove it with the sirloin muscles, but it can be scooped out for grind. There's no reason to let it go to waste!

Seaming the Bottom Sirloin Butt

The full sirloin can be seamed and separated into the subsections that make up the top and bottom butts. The top sirloin butt includes the top sirloin center and the sirloin cap; the bottom sirloin butt contains the ball tip and the tri-tip, which we will remove here separately. Sometimes the bottom sirloin will contain a portion of the sirloin flap; however, you already removed that muscle whole with the flank.

1. REMOVE THE BALL TIP.

The ball tip, a remnant of the sirloin tip after the sirloin and round are separated, can be removed through easily accessible seams. The three muscles – the *vastus medialis*, the *rectus femoris*, and the *vastus lateralis* – should be removed together; if large enough, save as a roast, otherwise toss into the grind pile or seam the muscles and cube for kabobs (see page 259).

2. SEPARATE THE TRI-TIP.

All that's left to remove of the bottom sirloin butt is the tri-tip (*tensor fasciae latae*). There's a straight seam between it and the top sirloin butt. Find that seam and separate the two muscles. For more information on the tri-tip, turn to page 280.

SIRLOIN BUTT

Seaming the Top Sirloin Butt

The top sirloin butt is the main source for sirloin cuts that are found in retail environments. The top sirloin is composed of two main muscles – *gluteus medius* and *biceps femoris* – and two minor muscles that are usually ground, *gluteus profundus* and *gluteus accessorius*. Roasts and steaks can be cut from the whole top sirloin butt or from seamed muscle groups. It's best to seam the top sirloin butt into the top sirloin center (*gluteus medius*) and sirloin cap (*biceps femoris*). The center will provide you with large or small sirloin roasts or steaks. The cap is actually the dorsal piece of the outside round, a muscle normally considered to be tough. This muscle is actually quite tender in the sirloin and can be cross-cut into portioned steaks or kept whole as a roast.

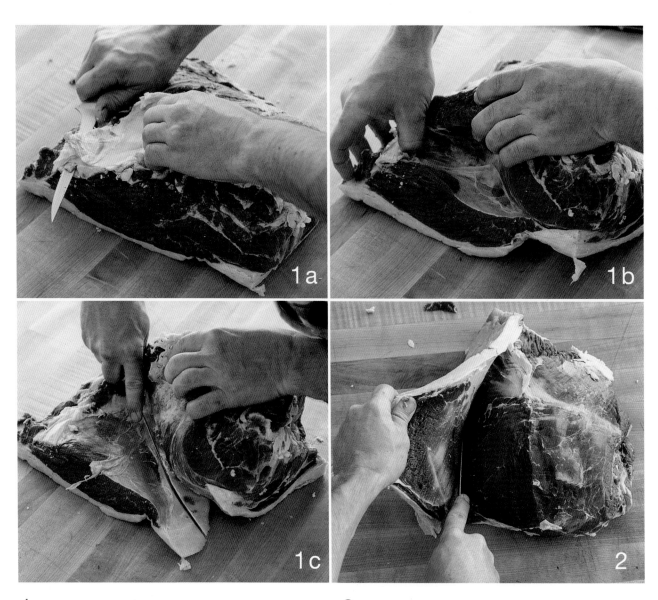

1. SEPARATE THE SIRLOIN CAP.
Trim away the fatty membrane that connected the aitchbone and sacrum. Separate the seam between the sirloin cap and top sirloin center, starting with your hand and finishing with the knife.

2. REMOVE THE FAT CAP.
The top sirloin center has a thick fat cap with some underlying connective tissue. Take both off at the same time.

3. REMOVE THE ACCESSORY MUSCLES.

First up is the *gluteus profundus*, the smaller of the two accessory muscles. Find its seam and remove it. Next is the *gluteus accessorius*, sometimes known as the mouse muscle. Its seam is easy to access after the other accessory muscle is removed. Open the seam with your fingers and then follow it with your knife. These smaller mouse muscles can be denuded and used as stew meat, or thrown into grind pile.

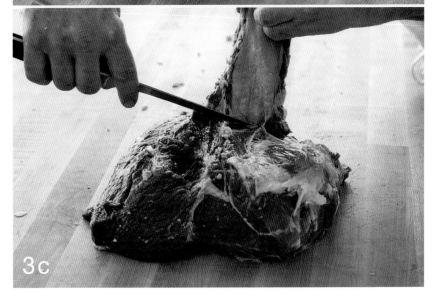

SEAMING THE TOP SIRLOIN BUTT

Breaking Down the Round

The round can be further dissected into its subprimals: top round, bottom round, and sirloin tip. There are many approaches to the order in which round subprimals can be removed. Here we remove them in the following order: the top round, the bottom round muscles (eye of round, outside round), and then the sirloin tip. The seams of the top round are the easiest to access initially; after its removal the remaining seams become much more straightforward.

Remove the Oyster and Aitchbone

The first step when breaking down a round is to remove the oyster steak and aitchbone. The oyster steak is a flat muscle that sits inside and attaches to the aitchbone. It is a small muscle – less than a single average serving – and more often than not is discarded due to its small size and proximity to where the bung was removed. However, you don't have to discard it: the oyster has an interesting texture that, when prepared correctly, is completely unique.

Start by removing the oyster. Use a semi-flexible boning knife and follow the interior contour of the aitchbone. The oyster has muscle grain that spreads and fans away from a central connective point. Work around this point as you release the muscle from the bone. Complete the removal of the oyster steak by severing the tough connective tissue that anchors it to the muscles on the opposite side of the bone. Trim away the exterior layer of the oyster if you plan on saving it.

1. FIND AND SEVER THE FEMUR TENDON.
Just interior to where the oyster steak sat is where the femur connects. Pull down on the aitchbone while tracing out the anterior edges. Keep your knife against the bone until you reach the femur. Sever the connective tissue around the ball-and-socket joint, pull down on the aitchbone, and use the tip of your knife to sever the tendon at the center of the ball-and-socket joint.

2. FOLLOW THE CONTOURS AND AVOID THE BUMP.
When the tendon is severed you will feel the aitchbone release, quite dramatically. Finish removing the bone by following its changing contours, working your way from anterior to the posterior side. On the posterior edge of the aitchbone is a considerable bump that will present an interesting challenge. Trace the bump with your knife in an effort to minimize the amount of meat left on the bone.

ROUND

Separate the Top Round

The subprimals are, more often than not, based on natural seams. For all round subprimals, a boning hook will help you apply outward force. This force further exposes the natural seams, which serve as an aid in showing you where to cut. Start with the top round, which is given its name because when the round primal is on a table, the top round naturally sits on top. Another name for this cut is the inside round because the muscles that make up this subprimal are on the inside of the leg.

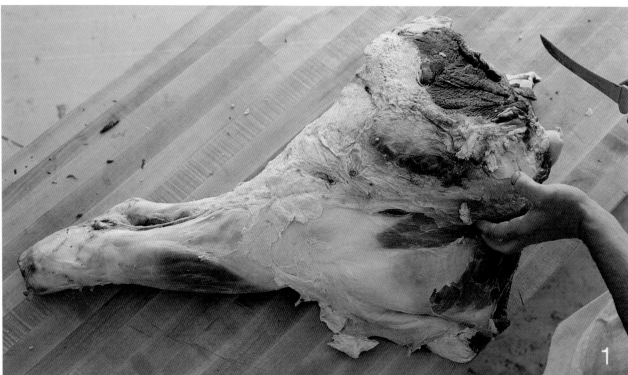

1. IDENTIFY THE TOP ROUND.
One edge of the top round sits next to the sirloin tip, and the other edge sits adjacent to the eye of round. The top is a thin tapered area that lies atop a portion of the heel. You will be separating the top round from these three adjoining muscle groups. Identifying the top round involves exploration, so take your time and work carefully. The heel is the least valuable of the three adjacent muscle groups, so do not fret if you slice into it while finding the seam.

2. CUT ALONG THE SEAM.
Start by separating the edge next to the sirloin tip. Use small exploratory cuts at first, keeping the blade between muscle groups while pulling the top round away from the sirloin tip. When that edge is largely released, follow the contour of the top round toward the heel and continue to follow the seam.

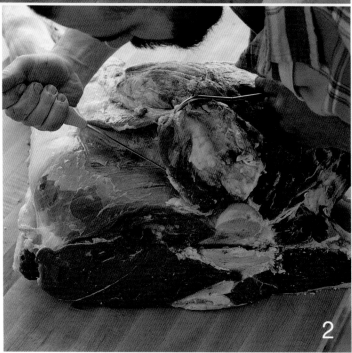

BEEF BUTCHERING

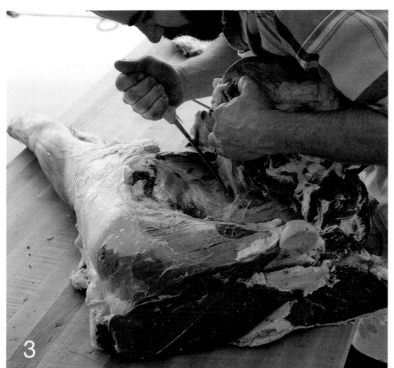

3. PEEL THE TOP ROUND.

Once the portion of the top round that sits atop the heel is released, peel the top round outward by severing connections near the interior of the round. Here you will find a grouping of connecting tissue and blood vessels. Severing these connections allows you to peel the top round away from the leg and will give you access to the eye of round from the inside.

4. FINISH THE SEPARATION.

Finish separating the top round by cutting along the flat edge of the eye of round.

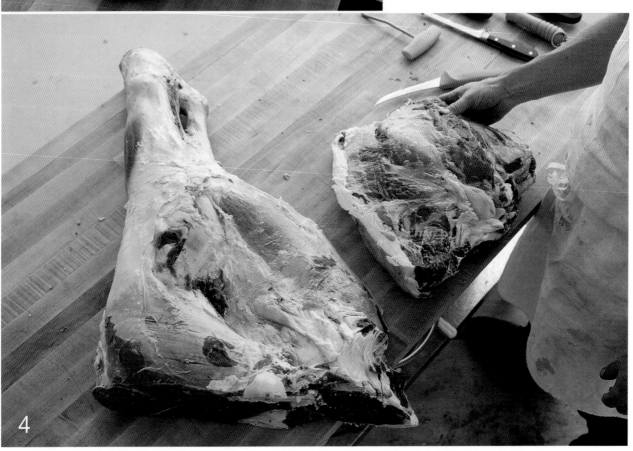

ROUND

Separate the Muscles of the Bottom Round

The bottom round, which is composed of the eye of round, outside round, and heel, can be removed as one whole subprimal. To save steps, separately remove each muscle group now, rather than later.

EYE OF ROUND

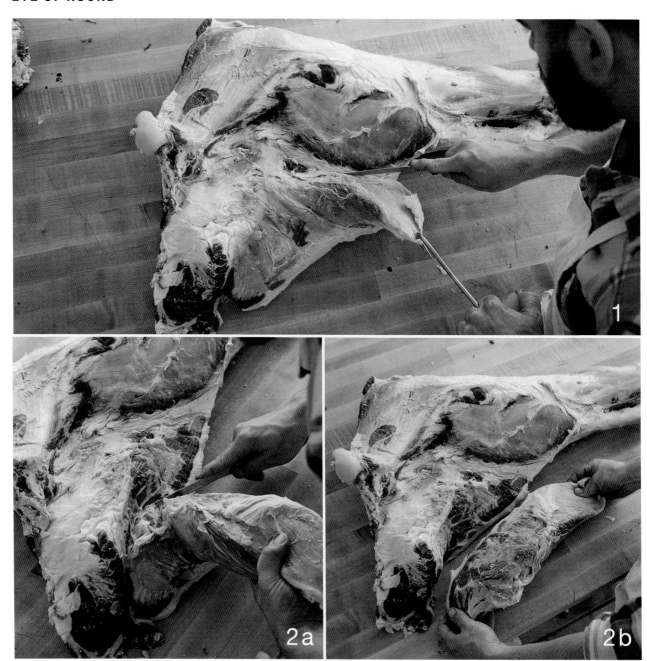

1. CUT ALONG THE SEAM.
The eye of round tapers on both ends but most dramatically near the heel, so start shallow with this seam. Find the top edge of the eye of round and carefully peel it away from the heel.

2. SEPARATE THE EYE OF ROUND.
Pull the eye of round away from the outside round, and then trace the seam between the two, working your way down in one smooth motion.

BEEF BUTCHERING

OUTSIDE ROUND

The outside round is also given its name because of its position on the table: it sits on the outside of the bottom round. Reposition the round as you wish to gain better access to the outside round. (As the image illustrates, the round is flipped over.)

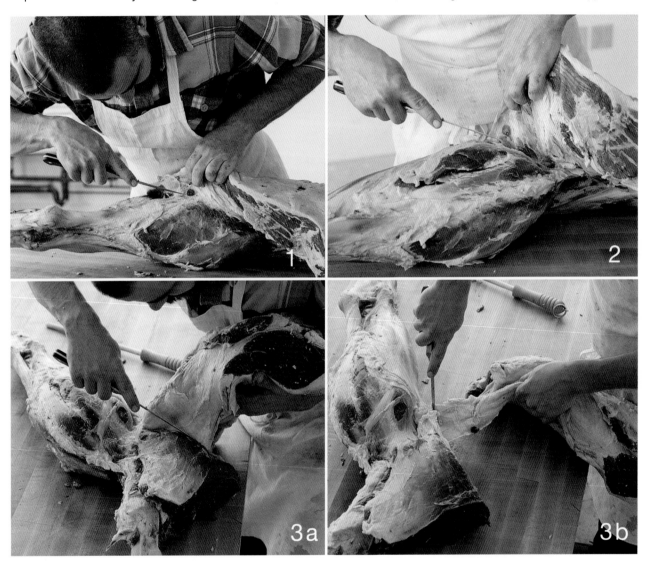

1. RELEASE THE TOP EDGE.
The top edge, like that of the top round and eye of round, is thin and lies atop the heel. Find it and follow the contour of the heel.

2. FOLLOW THE CONTOURS.
Continue down the heel until you reach a thick layer of silverskin. Sever it and pull the outside round away from the femur.

3. PEEL AWAY THE MEMBRANE.
Sever any connections between the outside round and the ball end of the femur. Follow the seam between the outside round and sirloin tip. Allow the membrane of the sirloin tip to stay connected to the outside round, allowing you to peel the membrane off the sirloin tip as you finish removing the outside round.

ROUND

Peel the Sirloin Tip

After completing the preceding steps, the sirloin tip is now the only remaining subprimal that is attached to the femur. You can see how the muscle tapers as it literally wraps around the bone.

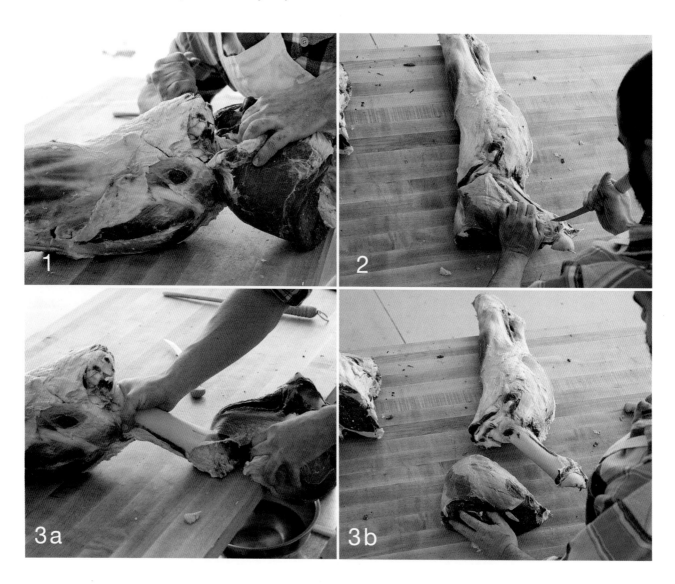

1. CUT BELOW THE PATELLA.
Just below the stifle joint (knee) is the patella (kneecap). The patella may be difficult to identify because the joint is covered with fat and connective tissue, but essentially you will be cutting where the fat stops and the connective tissue of the sirloin tip begins. Find that point and cut across it, down to the femur, making sure to score the bone sheath.

2. SCORE THE SIDES.
To peel the sirloin tip away from the femur you must first score both sides where the muscle ends and the bone sheath begins. Cut one side, making sure the sheath is fully scored, then repeat on the other side.

3. PEEL THE TIP.
Starting near the patella, pull the sirloin tip away from the femur, peeling the underlying membrane with it. Cut through the connections at the ball end of the bone to complete the removal.

ROUND

Finishing the Round: The Heel and Shank

Now that the subprimals are separated, we're left with the heel and shank, both of which are heavily worked muscles chockfull of collagen. The heel can be dealt with in any of the following ways: seamed for a surprisingly tender internal cut known as the merlot cut; left attached to the shank, enlarging the shanks offering; or removed and ground. Here we will remove it, giving you the option to seam it or grind it. The leftover shank is typically either cross-cut or boned and ground. Both fabrications are covered starting on page 298 of this chapter.

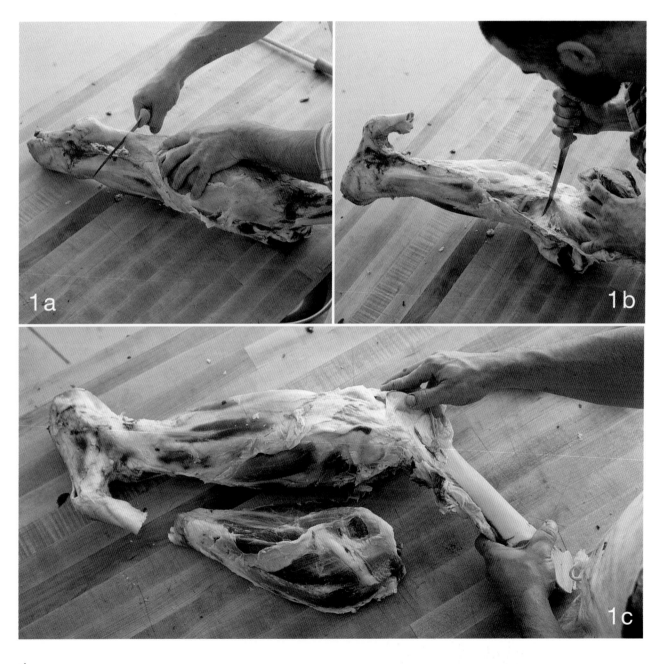

1. PEEL THE HEEL.

Sever the gambrel cord near the muscle, and then peel the heel away from the shank. Continue past the stifle joint and away from the femur, severing any remaining connections.

FINISHING THE ROUND: THE HEEL AND SHANK

ROUND

Finishing the Round: The Heel and Shank CONTINUED

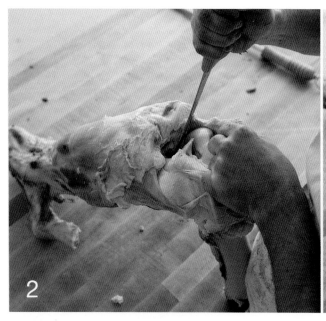

2. CUT THROUGH THE STIFLE JOINT.
Separate the femur by cutting around the middle of the stifle joint, through the connective tissue and fat, and exposing the interior. Inside the stifle joint is a thick ligament. Use the tip of your knife to sever it. Be careful not to bend your knife or use force in place of finesse.

3. SEVER REMAINING CONNECTIONS.
After the tendon is severed, continue cutting any remaining connections to fully separate the femur. Save this large bone for marrow bones. The shank can be processed further for cross-cut shanks, or can be boned for grind.

BREAKING DOWN A VEAL HINDQUARTER

The hindquarter of both veal and beef contains the same primals and sub-primals. Therefore, the breakdown for veal largely follows the same approach as that of beef, except that for veal the sirloin is left with the leg.

FULL TENDERLOIN REMOVAL. Veal, like beef, can have the entire tenderloin removed prior to separating the loin and leg primals. To do so, follow the instructions on page 195. Keep in mind that removing the tenderloin prevents you from cutting traditional loin chops, the veal equivalent of a beef porterhouse, because the loin chop includes a portion of the tenderloin. If you remove the entire tenderloin, you can cut chops that are the equivalent to a beef strip steak, or you can bone the loin to produce boneless roasts.

LOIN-LEG SPLIT. The hindquarter of veal is processed almost exactly like that of beef, with one exception: you should leave the sirloin with the leg rather than with the loin. In order to split the veal loin from the leg, refer to the beef instructions for separating the short loin from the sirloin on page 197. Further processing instructions for the leg start on page 285.

ADDITIONAL PROCESSING

THE PREVIOUS SECTIONS PROVIDE step-by-step instructions for breaking down a full carcass into quarters, primals, subprimals, and, in some instances, individual muscles. (Individual muscles are noted in italics adjacent to the cut names.) The remainder of this chapter focuses on additional processing for each primal, starting at the chuck and working toward the rear of the animal. Within each primal section, the cuts are organized alphabetically.

BEEF SUBPRIMALS & CUTS

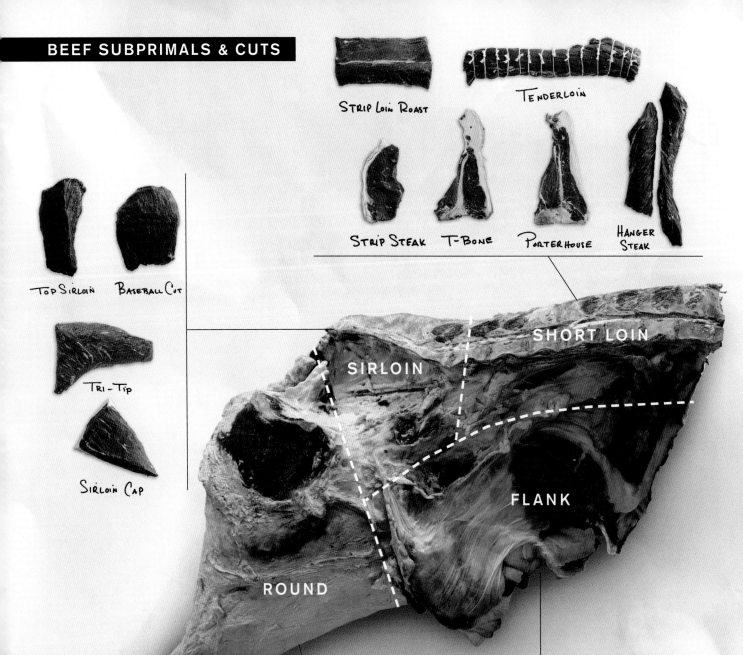

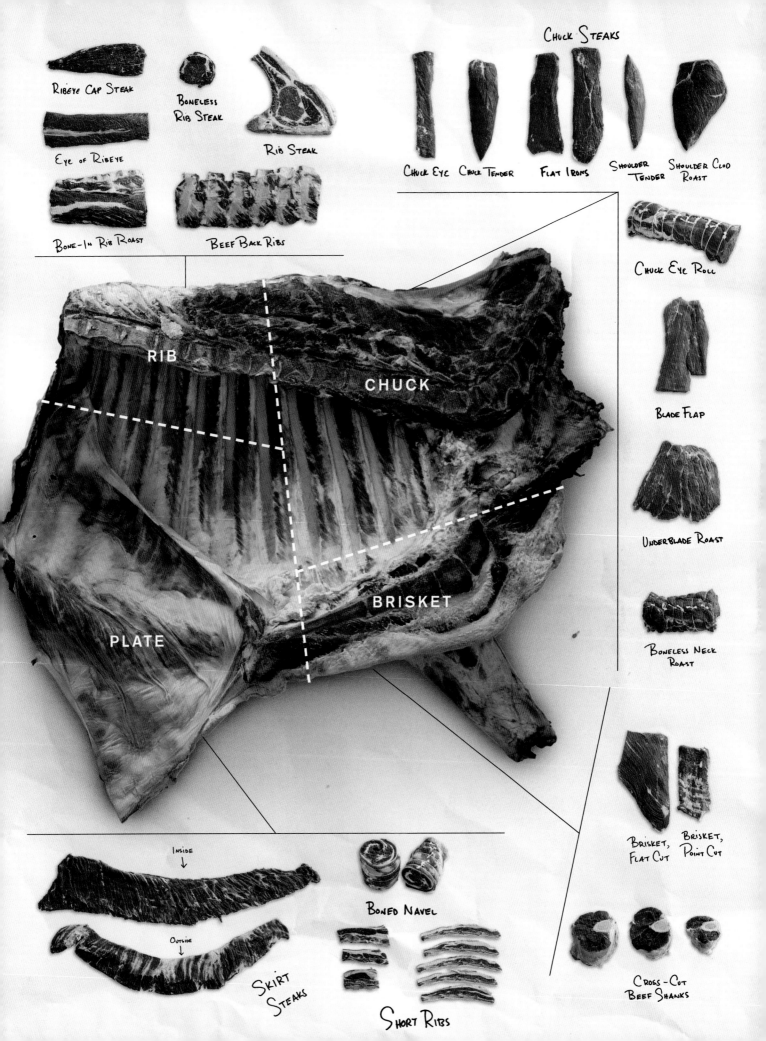

CHUCK

A VAST ARRAY OF CUTS come from the chuck, more than in any other primal. Here we cover each of the cuts that were separated while seaming out the chuck. I highly encourage you to explore the individual muscle cuts, learning more about their unique characteristics while gaining more value and diversity from the largest beef primal.

INCLUDED CUTS: blade flap, chuck eye, chuck eye roll, chuck tender, clod heart, neck, short ribs, shoulder tender, top blade, under blade roast

ADDITIONAL PROCESSING

Blade Flap *subscapularis*

The blade flap sits beneath the scapula. It is an intermittent working muscle with some intertwined connective tissue. One of the more tender muscles in the whole carcass, it makes a delicious thin steak once cleaned.

ALTERNATE NAMES: Bottom blade flap, under blade flap, Vegas strip steak

DENUDING: Remove any bone sheath, which is the fascial layer that was peeled off the bone. Flip the steak over; notice the various sections of silverskin. Remove as much silverskin as you can while maintaining the shape of the steak. Finish by trimming edges, which allows for even cooking.

Chuck Eye *complexus*

PORTIONING: After removal of the chuck eye from the chuck eye roll (see page 180), trim away any connective tissue. Cross-cut into portioned steaks, or tie every 2 inches to produce a small roast. When tying the full length into a roast, consider folding the tapered end for even cooking. (This process is similar to the tenderloin preparation on page 272).

Chuck Eye Roll

PORTIONING: Trim away any connective tissue from the exterior, but leave fat deposits. Begin tying every 2 to 3 inches, starting 1 inch from the end. If you want to produce steaks, cut between ties, making for 1- to 1½-inch steaks. For a roast, make a second pass of tying.

SEAMED MUSCLE PREPARATIONS: Separate all four muscles following natural seams (see page 180). The portion of the ribeye cap (*spinalis dorsi*) can be prepared as noted on page 252. Treat the chuck eye (*complexus*) according to instructions on page 182. Both the *multifidus dorsi* and *longissimus dorsi* can be cut into portioned steaks or tied and cooked whole. Consider using a Jaccard (see page 46) to tenderize steaks made from the *longissimus dorsi*.

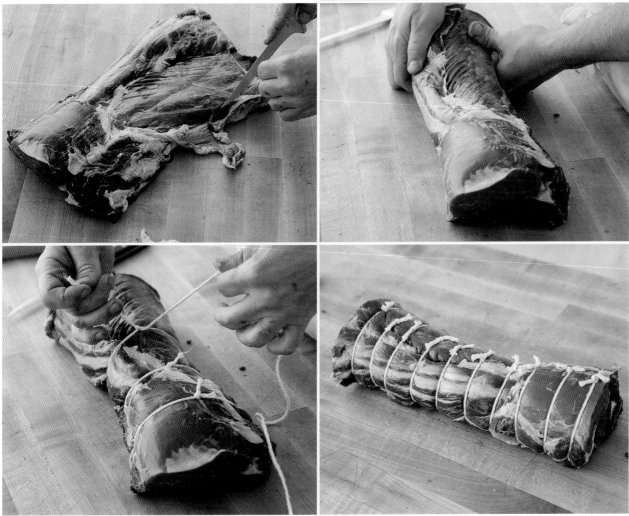

CHUCK EYE ROLL 219

Chuck Tender *supraspinatus*

ALTERNATE NAMES: mock tender, scotch tender, chuck filet

DENUDING: Trim away the exterior silverskin. Starting at the larger end and working with the grain direction, butterfly the muscle while following the thick layer of interior silverskin. Cut only a couple of inches into the muscle. Remove the silverskin following the underside as it tapers.

PORTIONING: Tie every 1 inch and cross-cut between ties for medallions; tenderize with a Jaccard. Tie every 2 inches for a roast.

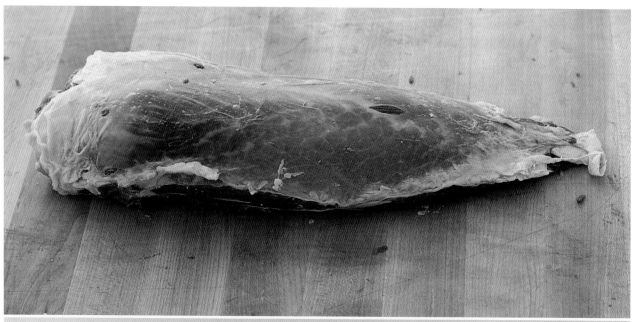

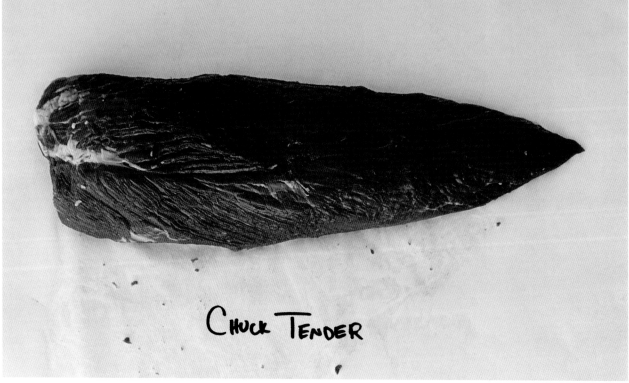

Chuck Tender

Clod Heart *triceps brachi*

ALTERNATE NAMES: chuck shoulder, shoulder clod, shoulder center, shoulder top, ranch steak

The clod heart of the chuck consists mainly of a single muscle group, *triceps brachii*, that is separated into three "heads": the long head, the lateral head, and the medial head. The long head makes up the bulk of this muscle, while the lateral head can also provide some portions. The medial head is a minor muscle, as is *tensor fasciae antebrachii*, both of which are removed and ground. The clod heart can be left whole and roasted, or trimmed down to the long head and cut into steaks or treated as a smaller roast.

CHUCK

Denuding the Clod Heart

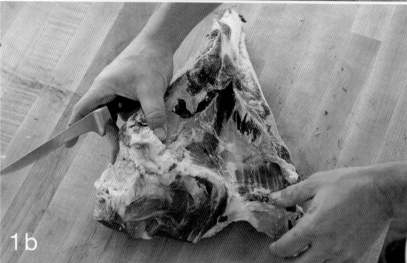

1. TRIM AWAY THE SMALLER SIDE MUSCLES. Start with the *triceps brachii* (medial head) and then the *tensor fasciae antebrachii*. Follow the natural seams, and save them for grind.

2. TRIM THE EXTERIOR.

There is a thick layer of silverskin that sits atop the larger of the remaining muscles, *triceps brachii* (long head), and wraps around the entire muscle group. Remove it, following the sloping shape of the muscles.

3. ROAST IT OR CUT STEAKS.

You can roast the remaining muscle group whole, or you can portion it to produce steaks that will need to be tenderized.

After denuding the clod heart you have the option of cutting ranch steaks, a portioned cut from just the *triceps brachii* (long head). The lateral head sits on top of this larger muscle and a thick sheet of silverskin; you can separate it by following the silverskin. Keep your knife on top of the silverskin and as it thins out, continue cutting straight to maintain an even shape of the underlying muscle. If the resulting lateral head is thick enough, cut steaks and tenderize them; otherwise save the meat for grind. Continue cleaning the long head by removing the exposed portion of the silverskin, starting at the larger end of the muscle. Remove any remnants of the arm tendon. For ranch steaks, cut across the grain and make them at least ¾ inch thick; tenderize using a Jaccard. For a roast, trim the tapered end of the muscle for even cooking.

Neck

The neck is an area that is very susceptible to contamination during the slaughter and aging process. Before any preparation, remove the first cervical vertebra and trim away any meat from the anterior end that appears to be dried, clotted, or unpalatable.

BONE-IN ROAST: Start with a bone-in neck. Trim away any excess connective tissue, glands, or remnants of the yellow backstrap (*supraspinous ligament*). Tie the roast in 3-inch sections to maintain shape through cooking.

CROSS-CUT: Follow the instructions for the bone-in roast, but instead of tying, cross-cut every 2 to 3 inches. Make sure to clean away any bone dust before cooking.

BONELESS ROAST: Follow instructions on page 174 for neck boning. Trim away any excess connective tissue, glands, or remnants of yellow backstrap (*supraspinous ligament*). Roll boneless meat, starting from the bottom (ventral) edge, and tie every 2 to 3 inches to maintain shape. Trim front and back to flat faces.

Short Ribs, Chuck *serratus ventralis*

ALTERNATE NAMES: chuck ribs, English ribs, flanken ribs, Korean-style ribs

The *serratus ventralis* is a large muscle, originating in the chuck near the second rib, and extends down to the twelfth rib. The loin (posterior) end is the more tender portion, but the chuck (anterior) portion also produces fantastic products. See page 239 for instructions on preparing the various short rib cuts. When boned (see page 231), chuck ribs provide some of the best stew meat in the entire carcass. The meat between the bones, or intercostal meat, can be removed, stacked, and tied for a riblet roast.

Shoulder Tender *teres major*

ALTERNATE NAMES: petite tender

The shoulder tender is another chuck muscle that ranks near the top of tender muscles in the entire carcass. It is a small muscle and weighs about 1½ pounds. There are only two shoulder tenders per carcass; hence, the per-animal yield is small.

DENUDING: Trim away any exterior silverskin that is left after separation from the other shoulder clod muscles, working in the muscle grain direction.

PORTIONING: Roast whole, or cross-cut into medallions that are 1 inch thick or thicker.

ADDITIONAL PROCESSING

Top Blade *infraspinatus*

ALTERNATE NAMES: flat iron steak, top blade steak, chicken steak

The *infraspinatus* is the second-most-tender muscle in the entire carcass, just after the tenderloin. This tenderness is due to a large swath of connective tissue that runs right through the middle of the muscle. This connective tissue provides so much support for the animal that the two sides of the muscles don't need to do much work. The top blade is also one of the more difficult muscles to denude, due to said layer of connective tissue. However, the end result, the popular flat iron steak, is worth the effort.

CHUCK

Making Flat Iron Steaks

1. **FIND THE CONNECTIVE TISSUE.**
The top blade tapers as it flattens out. That wandering sheet of connective tissue begins at the thicker (ventral) end of the muscle, nearest the joint end of the scapula. Place the top blade fat-side down (bone sheath up) and cut into the thicker end to find the top of the connective tissue. Stay flat with the connective tissue and follow it to the edge, where you will continue separating it from the overlying muscle.

2. **FOLLOW THE EDGE AND WORK ACROSS.**
Continue working down the side of the muscle, staying parallel to the connective tissue. As you progress down the muscle, work your way across as well as down by making shallow strokes while folding the muscle back. The connective tissue sheath is wavy, making it even more challenging to follow. Proceed slowly, making exploratory cuts, until you get the hang of it. Follow the connective tissue to the end of the muscles, separating the two sides.

ADDITIONAL PROCESSING

3. DENUDE THE TWO SIDES.

The (medial) side that you just removed will still have the bone sheath on it; the other (lateral) side will have connective tissue on both sides. Both need to be denuded. You can use the same denuding method for all sides. In preparation for removing the bottom layer of connective tissue, place your hand on the steak to keep it flat against the table. Cut into the muscle at one end and get your knife flat against the connective tissue and parallel to the table. Hold down that end and then cut the entire length of the muscle, keeping your blade parallel with the table. Repeat on the all sides that need cleaning. Trim up the edges of both steaks and remove any lingering silverskin.

CHUCK

Cutting Top Blade Steaks

Top blade steaks, also known as chicken steaks, are simply a cross-cut top blade steak, trimmed of exterior fascia but with the interior silverskin left on. Clean the exterior layers of silverskin according to preceding directions. Cross-cut steaks every 1 inch, or thicker, and make sure to tenderize them with a Jaccard.

Under Blade Roast

ALTERNATE NAMES: Denver roast, boneless chuck rib roast

The under blade roast is composed of three muscles: *rhomboideus*, *splenius*, and *serratus ventralis*, with the latter being the main muscle. Previously, during the separation of the under blade, you removed the *rhomboideus* (see page 179), which left the two other muscles. These two muscles can themselves be separated using natural seams.

SEAMING: The *splenius* is a flat muscle that sits atop the outer (lateral) face of the *serratus ventralis*. It's easily removed by following an underlying seam. Cleaned, the *splenius* is known as the Sierra cut; when cross-cut into steaks, it's known as the Sierra steak.

DENUDING: Trim any thick exterior fascial layers or other unpalatable connective tissue. Trim tapered edges to promote even cooking.

CUTTING DENVER STEAKS: Start at the thicker (posterior) end and cut steaks crosswise. Make each steak about 1 inch thick.

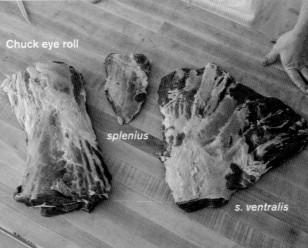

A cleaned *serratus ventralis*.

BRISKET

THE BRISKET IS COMPOSED OF two muscles: the deep pectoral (*pectoralis profundi*) and the superficial pectoral (*pectoralis superficialis*). The former is known as the flat cut, while the latter is known as the point cut. The names are derived from the shape and position of each when the brisket is whole. Brisket can be prepared bone-in (if removed in the formation of the square-cut chuck); boneless and whole; or boneless, seamed, and trimmed to lean muscle. For processing information on the foreshank, turn to page 159.)

INCLUDED CUTS: brisket flat, brisket point

BRISKET

Boning the Brisket

The brisket is most commonly boned, as we show here, allowing you to prepare it whole or separate it into its two parts: the flat and the point.

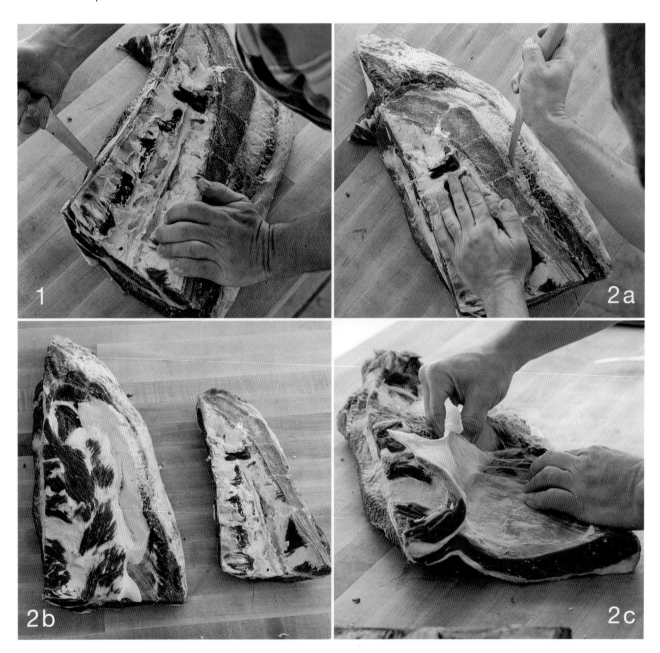

1. FOLLOW THE FLAT OF THE RIBS.

Separate the underside of the ribs, starting at the posterior edge where the plate was separated. Stay close to the bone and don't cut farther in than the sternum.

2. BONE THE STERNUM.

Trace the sternum, starting at the point, cutting deep enough to reach the natural seam that lies beneath the sternal fat. Pull up on the bone and sever any remaining connections. Remove the excess sternal fat and rib meat through the underlying seam. Save this excess for grind.

BRISKET

Seaming the Brisket

Seaming the brisket enables you to work with small cuts, as roasting or brining a whole brisket may be more than what you're in the market for. It also allows you to remove unwanted fat deposits. (Save the fat, though, and add it to your ground beef.)

SEPARATE THE MUSCLES. Lay the brisket on the table with the boned surface facing down. The muscle on the bottom is the flat (deep pectoral); the muscle on the top is the point (superficial). Follow the upper surface of the deep pectoral, starting at the anterior end. Trim away the exterior fat on both muscles to a desired level.

ADDITIONAL PROCESSING

PLATE

THE PLATE INCLUDES TWO of my favorite cuts: the outside skirt and a nice, large portion of the *serratus ventralis*, the main muscle in short ribs. Where you separated the plate from the rib primal will determine the length and volume of short ribs you'll have; I opt for more short ribs and a shorter tail on my rib steaks. As you finish off processing this primal, don't overlook the navel as a stand-alone cut, which offers a great option for pastrami.

INCLUDED CUTS: inside skirt, navel, short ribs, outside skirt

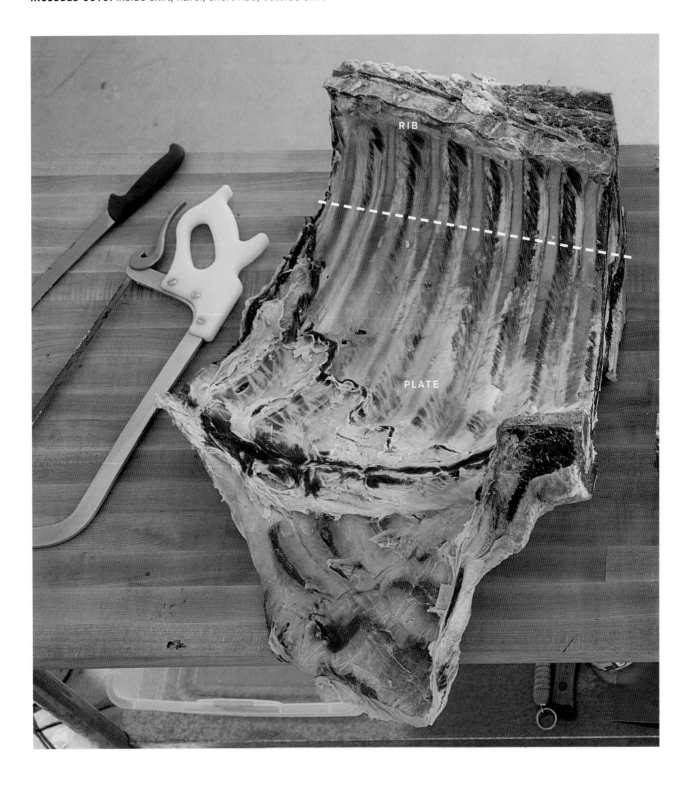

Navel

Boning the upper (dorsal) portion of the plate provides you with boneless short ribs, which are best prepared when the fat and muscle are trimmed down to the *serratus ventralis* muscle. The lower (ventral) portion of the plate, from the rib tips and costal cartilage down, is considered the navel — a cut that is best when boned, as shown in this sequence.

1. **REMOVE THE BONES.** Place the bone-in navel bone-side up. Note the shape of the costal cartilage and the bottom (ventral) edge. Separate the cartilage first and then work your way down the ribs, staying close to the bones while using long strokes. The cartilage can be easily cut through with a knife, so be careful not to leave portions of it behind.

2. TRIM IT UP.
Square up the sides of the navel. Remove the intercostal meat and underlying connective tissue. The result is a clean, flat surface.

Boned & Rolled Navel

BEEF RIB CUTS

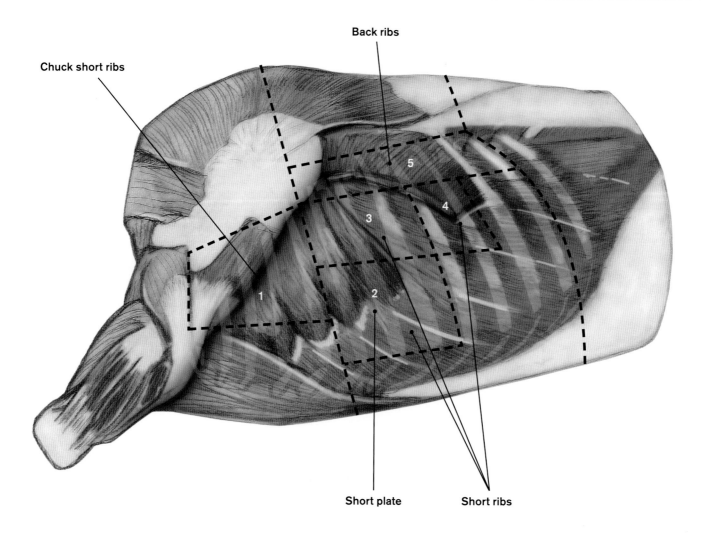

Chuck short ribs
Back ribs
Short plate
Short ribs

THE MUSCLE profiled in this diagram, *serratus ventralis*, defines the areas where short ribs can be cut from. It originates in the chuck, around rib 2, and, as you can see, pretty much covers the whole area of **chuck short ribs** (area 1), making them one of the best sources for short ribs. The muscle is expansive and thick in this area.

AS THE MUSCLE spreads rearward it tapers in thickness. Areas 2, 3, and 4 provide the more traditional **short ribs** that come from the plate primal. Note that while the muscle spans these three sections, area 2 is the only one with full coverage. This area, provided as one slab, is commonly known as the **short plate**.

THE REAR of the *s. ventralis* muscle, around ribs 9–12, has tapered to too thin a thickness to cut bona fide short ribs. This area is better boned and rolled or added to grind.

SITTING ATOP *s. ventralis*, separated by a layer of fat, is *longissimus dorsi*, a muscle that also spans much of the upper thoracic rib area. It is these two muscles that provide the meat to short ribs, with *s. ventralis* being the preferred of the two.

THE DORSAL PORTION of ribs 6–12, once removed from the ribeye roll and chine bone, are known as **back ribs** (area 5). They primarily include the intercostal meat between the ribs.

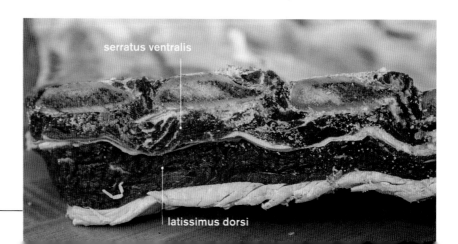

serratus ventralis
latissimus dorsi

ADDITIONAL PROCESSING

Short Ribs

ALTERNATE NAMES: English ribs, flanken ribs, Korean-style ribs

Short ribs can be prepared as a whole slab, but more commonly they are cut in a few other ways. English-style cuts are prepared by separating the ribs lengthwise into single bones after they have been cross-cut to 3 to 6 inches in length. Flanken-style ribs are cross-cut sections of two to three ribs. Korean-style ribs are cut to ¼ inch in length, while 1- to 2-inch cuts are more common for American cuisine.

ENGLISH STYLE

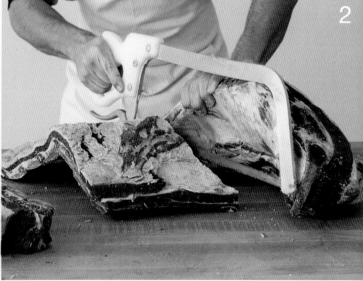

1. MARK YOUR CUT LINE.
Choose how long you want the ribs. Cut accordingly, down to the bone, while staying parallel to the top (dorsal) edge.

2. SAW DOWN THE LINE.
Separate the section of ribs by sawing along the cut line.

3. SEPARATE THE RIBS.
Cut in-between each rib.

The thicker the *serratus ventralis* muscle the better. The ribs at left are good; those at right are not so good.

Cutting Short Ribs CONTINUED

FLANKEN STYLE

The thicker the *serratus ventralis* muscle the better. The rib on the right is good; the rib on the left is not so good.

1. MARK YOUR CUT LINE.
Choose how wide you want the ribs. Cut accordingly, down to the bone, while staying parallel to the top (dorsal) edge.

2. SAW DOWN THE LINE.
Separate the section of ribs by sawing along the cut line.

3. SEPARATE THE RIBS.
Separate the row of ribs into two- or three-rib sections.

ADDITIONAL PROCESSING

Skirt Steak, Inside *transversus abdominis*

ALTERNATE NAME: skirt steak, fajita meat, flap meat

The inside skirt extends across the flank and plate primals and, unless removed from an entire hanging side, is removed in segments from said primals. It has a similar texture and graininess to the outside skirt steak, but due to its exterior location it does not share in the deeper, organ-influenced palette of the outside skirt.

CLEANING: The inside skirt needs very minimal cleaning. Remove any exterior connective tissue or membranes, but leave the fat deposits.

Skirt Steak, Outside *diaphragm*

ALTERNATE NAMES: skirt steak, fajita steak, Philadelphia steak

The outside skirt is oddly named, considering that it resides on the inside of the animal's abdominal cavity (by contrast, the inside skirt sits on the outer edge). The outside skirt is a cut that was popularized in the early 1980s, when it was marketed as a fajita-style steak. The outside skirt is a portion of the diaphragm, a muscle that is frequently used by the animal (read: deep flavor) due to its role in breathing, while its thin, fibrous structure gives it a unique texture.

DENUDING: A thick membrane covers both sides of the outside skirt. This membrane can be easily peeled away after the edge seams are removed.

Because the membranes on both sides converge at the edges, it's easier to remove the membranes after those points are removed. Trim away ¼ to ½ inch from both edges of the outside skirt. Make sure to trim away enough so that you can remove the seams of connection tissue.

Use one hand to keep the muscle flat against the table, and use your other hand to peel away the membrane. This method will help prevent tearing and damage to the steak, an easy occurrence due to the loose structure of the diaphragm muscle fibers. Start peeling the membrane from either end while using a flat palm to keep the exposed meat flat on the table. Continue moving your palm to the edge of the membrane while peeling with the other hand until the membrane is completely removed. You may run into small areas that require knife work in order to complete the separation.

The exposed faces of the diaphragm have another thin layer of connective tissue that needs to be trimmed away. Using a long-bladed knife, trim the layer from the edges of the muscle. Leave any fat deposits for flavor and self-basting.

RIB

THE RIB IS THE ONLY PRIMAL THAT you can practically cook as is, but it behooves you to do some easy cleanup to make it really shine. And while a beef side produces many kinds of steaks, the rib steak is arguably the greatest: it includes more muscle groups and fat, which results in more diversity of flavors. (Keep the ones closer to the chuck if you're deciding to share your bounty.) For those interested in seaming this primal, don't forget about the unsung hero, the ribeye cap steak, a small muscle that's certainly worth the effort of isolating.

INCLUDED CUTS: bone-in rib roast, bone-in rib steak, bone-on rib roast, boneless rib roast, boneless ribeye steak, ribeye cap steak, ribeye roll

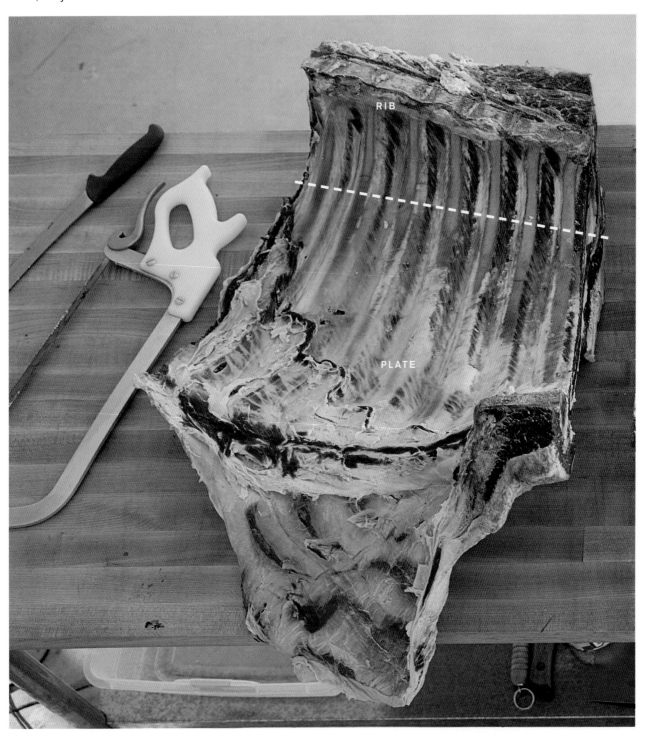

Bone-In Rib Roast

ALTERNATE NAMES: rib loin, standing rib roast, bone-on rib roast

Bone-in preparations of the rib can produce some of the more impressive presentations and cuts from a beef carcass. The following sections describe how to create an oven-ready bone-in rib roast from the primal, a process that involves three steps: removing the rib cover, chining the rib, and trimming the rib. This process will also set you up for additional options, like a bone-on roast or rib steaks. The measurements provided in these instructions are guides and, as with separating the rib and plate primals (see page 142), your personal preferences should play a role in where cuts are made.

The posterior remnants of the shoulder blade and the rib cover (a layer of fat and muscle that lies atop the more desirable muscles of the rib primal) must both be removed. The rib cover includes the following muscles: *latissimus dorsi*, *infraspinatus*, and *trapezius*, which are above the shoulder blade; and the *subscapularis* and *rhomboideus*, which are below the shoulder blade. The fat outside the *latissimus dorsi* (also known as the lifter meat) can be trimmed off as a whole sheet and then wrapped across the roast, which provides a covering for the roast and basting during cooking.

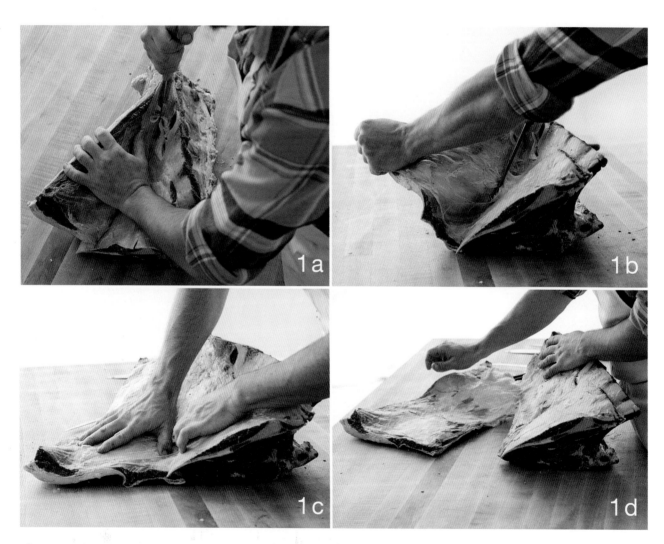

1. REMOVE THE RIB COVER.

Removing the first two muscles – *latissimus dorsi* and *trapezius* – can almost be done by hand. The seam is best accessed from the ventral edge (where the ribs were cut). Peel the thin muscles and fat away, following the seam, and remove them, using a knife when necessary. As you make your final cut, at the bottom edge, be careful not to cut into the underlying rib cap (*spinalis dorsi*).

2. REMOVE THE SCAPULA AND ATTACHED MUSCLES.

Like the lifter meat, the scapula can practically be peeled away by hand. (You can combine this step with the previous step, by the way.) Attached to the top of the scapula is a piece of the top blade (*infraspinatus*). Get underneath the scapula and the two smaller muscles that are attached, *subscapularis* and *rhomboideus*. Follow the seam, being careful not to damage the rib cap as you remove the *rhomboideus*.

3. CHINE THE RIB.

Removing the chine bone (an industry term for the spine) allows you to easily remove the feather bones (as known as the *spinous process*) while leaving the ribs on. The other benefit of removing it now is that, after cooking, the bone-in portions can be cut without the need to cut through the chine bone.

Saw through the base of the ribs at about a 45-degree angle. The saw blade should just barely graze the muscle that attaches to the spine, the *multifidus dorsi*, and then through the feather bones where they connect to the vertebrae. Line up your handsaw and saw through the ribs, holding the roast flat on the table or standing on end.

Bone-In Rib Roast CONTINUED

If you stop work after chining, you will have a nice standing rib roast. However, you can continue trimming, removing some of the other layers of connective tissue to pretty up the roast. Fattier beef may have a layer of fat on top of the exterior membrane. If this is the case, you may choose to leave it on. If so, just remove the backstrap (step 5), and skip steps 4 and 6 here.

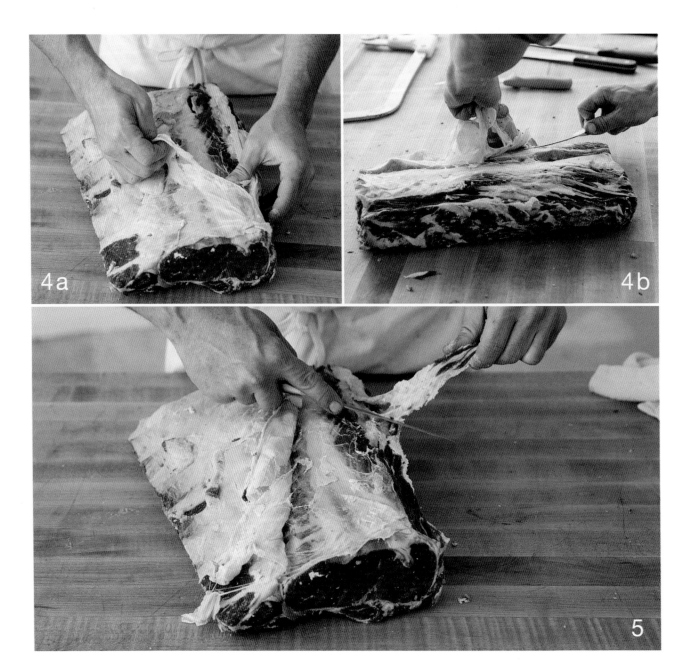

4. PEEL AWAY THE MEMBRANE.

The membrane that lies over the muscles is easily separated by hand. Be wary of pulling too hard and damaging the rib cap. Pull back the membrane, remove it from the muscles, and continue removing it from the ribs, taking a little fat with it.

5. REMOVE THE BACKSTRAP.

The thick, elastin ligament that runs along the bottom of the muscles is known as the backstrap. Get under it with your knife and pull it away to remove it. (Shown in the accompanying image, this step is being done halfway through the membrane removal process in the previous step.)

6. TRIM THE SILVERSKIN.

A swath of silverskin runs across the main loin muscle and underneath the rib cap. Although you will want to leave the rib cap on, it's best to denude the visible portion of the loin muscle. Start at the chuck end and work your way down the muscle, working with the grain. After that silverskin is removed, trim away the striated pieces of silverskin that sit atop the rib cap.

Bone-In Rib Roast

Bone-On Rib Roast

Bone-in rib roasts make for fantastic meals, but sometimes navigating the ribs after roasting can be a bother, either for you in the kitchen or for others on their plates. Because of this, a bone-on rib roast is ideal: you get the benefits of cooking meat on the bone, but after cooking the bones will fall away by simply cutting some strings. Start with a bone-in rib roast prepared according to the aforementioned instructions, starting with Removing the Rib Cover (see page 244). (As with a bone-in roast, a bone-on roast can be prepared with the lifter meat fat trimmed off as a whole sheet and then tied on so that it covers the rib roast during cooking.)

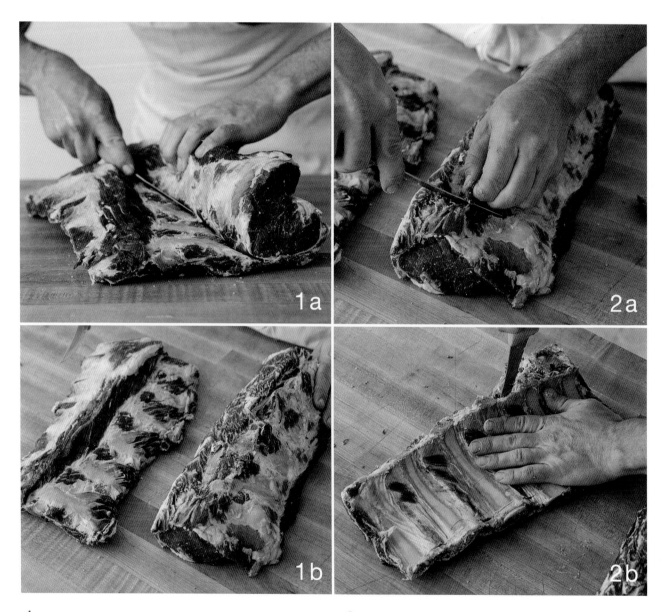

1. REMOVE THE LOIN MUSCLES.

Follow the flat of the rib bones with your knife, starting just above the eye of the ribeye, and remove with the ribeye cap. (Together these are known as the ribeye roll.) This leaves the tougher muscles on the bones.

2. CUT OUT ANY BONE BUTTONS.

At the base of the ribs you may find some bone buttons, remnants of the vertebrae that were left after chining. Scoop those out of the meat with your knife or separate them from the ribs.

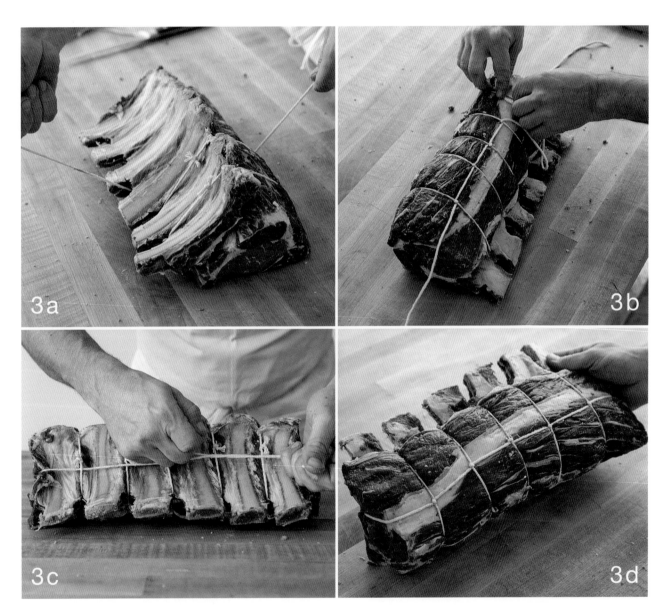

3. TIE THE BONES BACK ON.

Place the meat against the table and set the bones on top. Secure the bones with a piece of string between each rib. Then take a long piece of twine and, on the front of the roast, loop it around each of the ties. Bring both ends of the twine around back and weave it through the ties before making a final knot.

Rib Steaks, Bone-In

ALTERNATE NAMES: cowboy steak, scotch filet, ribeye steak, rib steak

In theory, bone-in rib steaks can be cut during any stage of the bone-in rib primal breakdown. But I highly recommend chining before cutting bone-in rib steaks because it makes the final product more attractive and is easier to eat on the plate. Plus, you can cut them just using a knife. Because there are seven bones, you should expect to get seven steaks. However, if you have a bandsaw, you can cut steaks based on a uniform measurement, no thinner than ¾ inch. Just remember to scrape bone dust afterward.

As shown in the accompanying images, cut bone-in steaks after trimming and chining the primal (see page 245). Use a long-bladed knife to cut steaks, and make each cut in one motion to prevent any serrations on the face of the steak. Cut each steak in between the rib bones. Place your free hand against the face of the meat to promote even and straight cutting.

ADDITIONAL PROCESSING

Rib Steaks, Boneless

ALTERNATE NAMES: Delmonico, scotch filet, boneless rib steak, ribeye roll steak

Boneless rib steaks are typically cut from the ribeye roll preparation. As with bone-in steaks, use a long-bladed knife. Cut the steaks in one motion to prevent serrations on the face. Keep one hand pressed against the face of the muscle to ensure a flat cut. Cut steaks ¾ inch thick or thicker.

Ribeye Cap Steak *spinalis dorsi*

The ribeye cap is arguably the best muscle in the rib primal, and only just recently has it been getting its due attention. Follow the instructions on page 256 for removal.

DENUDING: First, trim the edges. Then, denude it like you would any flat steak, keeping the meat flat against the table and your knife parallel with the surface. Cut from the tapered end toward the wider end. Portions should be cut against the grain.

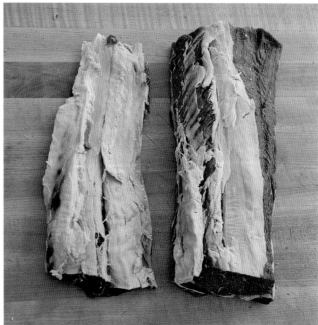

Ribeye Cap (Spinalis Dorsi)

252   ADDITIONAL PROCESSING

Ribeye Roll

ALTERNATE NAMES: boneless rib roast, boneless ribeye roast, boneless rib loin

Boning the rib primal provides a lot of possibilities, especially when this process is combined with the seaming out of muscle groups. The revered ribeye cap is obtained using this preparation method. The rib primal can be boned without the extra step of chining (as shown on page 245).

1. REMOVE THE RIB COVER.

Begin by removing the rib cover and muscles around the scapula, per the instructions on page 244. (Or, if you want to leave some or all of the rib cover and muscles on, you can do that, too.) As with the bone-in iterations, you can save the exterior fat, trim it down to a sheet, and tie it over the roast for some extra flavor and basting during cooking.

Ribeye Roll CONTINUED

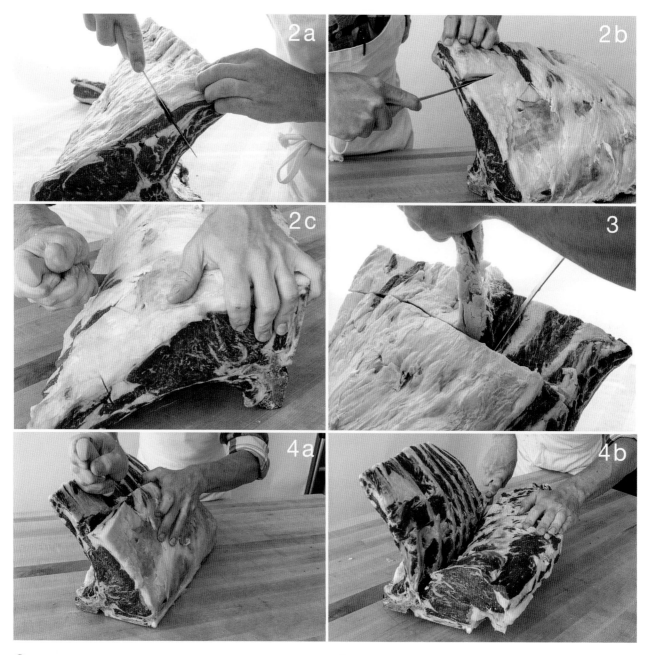

2. CUT ACROSS THE RIBS.

(You can trim this lip off of the roast before or after boning, though I prefer to do it first.) Find the top (ventral) edge of the main loin muscle and then, an inch or two above the loin muscle, cut across all the ribs, staying parallel to the loin muscle.

3. REMOVE THE LIP.

Follow the flat of the bones and remove the fatty lip from the roast, providing easier access to the ribs and ribeye roll.

4. BONE THE RIBS.

Bone the ribs from the loin muscle as one piece, staying flat against them as you follow their curved shape toward the vertebrae. At the base of the ribs is a bump of bone called the costovertebral joint. Turn your blade outward and cut, just an inch or so, to get past the joint. Once you've cleared the joint, angle the blade back toward the vertebrae and continue cutting until you reach the feather bones.

ADDITIONAL PROCESSING

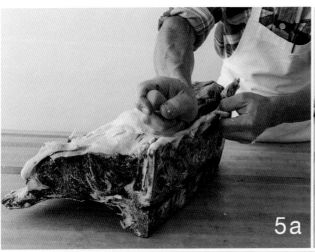

5. CUT ALONG THE FEATHER BONES.

Place the roast on end, ribs against the table. Cut along the cartilaginous tips of the feather bones and then follow their flat shape until you connect with the cut you made while boning the ribs, thus finishing the removal of the bones. (If you didn't trim the lip in steps 2 and 3, do so now.) This completes the procedure for creating a ribeye roll. You can roast the ribeye roll whole, portion it into smaller boneless roasts, or cut it into boneless rib steaks.

6. CUT BACK RIBS.

With the ribeye roll fully boned, you can easily create beef back ribs. Saw across the ribs at their base, as close to the chine bone as possible. Peel the membrane off the backside of the bones, and they are ready to barbecue. Save the chine and feather bones for stock.

RIB

Seaming a Ribeye Roll

There are two muscles in the rib primal worth extracting by themselves: the main eye of the ribeye (*longissimus dorsi*) and the delectable ribeye cap (*spinalis dorsi*). Bone and trim the rib to a ribeye roll before you begin.

1. SEPARATE THE CAP.

The tapering muscle that lies atop the eye is the ribeye cap. Approach it from its top (dorsal) edge, starting the seam with your knife and continuing to open it with some help from your hands. Follow the seam and then cut away the cap. (Further instructions for working with the ribeye cap are on page 252).

ADDITIONAL PROCESSING

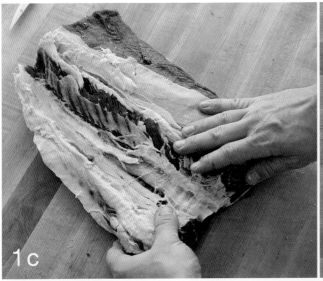

2. TRIM THE EYE.

The removal of the cap will reveal a very thick sheet of silverskin atop the eye. At first glance, the silverskin may appear to be just like any other curved surface. However, this one has a deep valley of connective tissue running down it. Don't bother trying to dig out the silverskin. Instead, keep your knife level with the rest of the muscle and cut across the valley. Stay shallow and work with the grain direction, which runs from the chuck (anterior) end back.

Eye of Ribeye (Longissimus Dorsi)

LOIN

THE LOIN PRIMAL PRESENTS YOU with the main crossroad in butchering: remove the tenderloin, or not. (I prefer to remove it and eat it raw, sliced paper thin as carpaccio.) In either case, the loin is providing some of the most tender cuts in the carcass: the tenderloin tops the bunch, and the loin section of the *longissimus* muscle is its more tender portion. The two subprimals are the short loin and sirloin, with the latter providing an array of muscles, many of which are worth isolating for their respective benefits.

INCLUDED CUTS: hanging tender, bone-in loin steaks (strip steaks, T-bone steaks, porterhouse steaks), bone-in loin roast, boneless loin roast, boneless loin steaks (strip steak), sirloin (bottom, top), sirloin cap, tenderloin, tri-tip

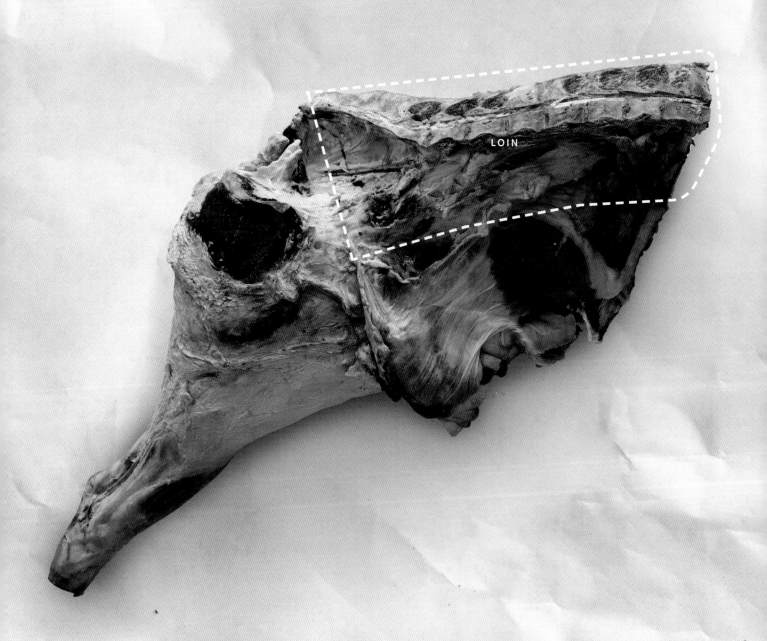

Ball Tip

ALTERNATE NAMES: ball tip roast, ball tip steak

The ball tip is the dorsal remnant of the sirloin tip, which is left on the sirloin after separation from the round primal. Three of the four sirloin tip muscles are in the ball tip: *vastus medialis*, *rectus femoris*, and *vastus lateralis*. If the ball tip is sizable enough you can roast it whole, cross-cut steaks from it, or seam the three muscles. The seams between the muscles are easily accessible: *vastus medialis*, the smaller side muscle, should be removed first; the remaining two muscles are joined by a seam of silverskin that can be followed to separate them and then denuded from either side.

All that's left to remove of the bottom sirloin butt is the tri-tip (*tensor fasciae latae*). There's a straight seam between it and the top sirloin butt. Find that seam and separate the two muscles. Clean up the tri-tip by denuding the underside and trimming the exterior fat to a desired level. Follow the instructions on page 202 for further instructions on seaming and cleaning the top sirloin butt.

Hanger Steak

ALTERNATE NAMES: butcher's steak

The hanger steak is the dorsal portion of the diaphragm, attaching to the spine near the last thoracic vertebra. This area of the diaphragm is also where the esophagus and two major blood vessels pass between the thoracic and abdominal cavities. (You can see a tubular portion of the artery in the photo below.)

Like the outside skirt steak (see page 242), the hanger steak is a very worked muscle (read deep flavor) with an evident grain; its proximity to the kidneys, and other viscera, also gives it a unique flavor. Before cooking it an exterior membrane needs to be removed, as does a thick swath of connective tissue that runs through the middle of it, resulting in two separate, unequally sized steaks.

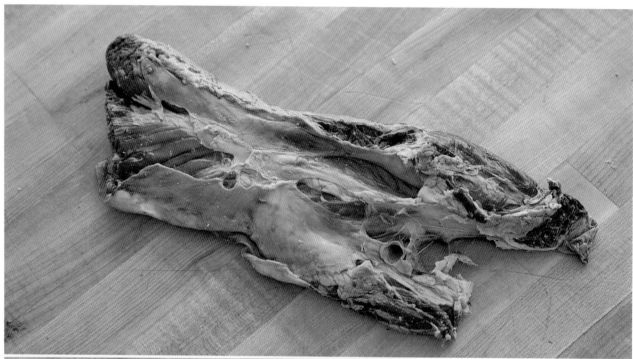

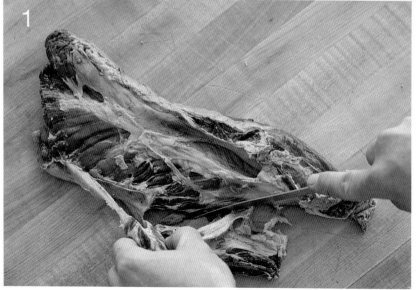

1. PEEL THE MUSCLE.
The hanging tender is covered in a membrane sheath, similar to that of the outside skirt steak. Get started by peeling away the membrane sheath with your fingers. The attachments on the side will need some knife work. Clean both sides down to the muscle.

2. SPLIT INTO TWO SIDES.

A thick band of connective tissue runs at an angle through the middle of the muscle. Check its location on both sides to understand the angle that you'll need to follow. Start on one side, following the face of the connective tissue, until that side of the muscle has been completely removed. Repeat with the other side. Finish with some final cleanup of the surfaces, leaving as much surface fat as you can.

Loin Steaks, Bone-In

The short loin provides some of the most well-known and desired cuts from a beef carcass. Whether to include or exclude the tenderloin in your bone-in preparations is one of the first decisions you must make when dealing with the short loin. When included, you'll be cutting porterhouse and T-bone steaks; when excluded, this subprimal becomes a strip loin (see page 265), which can produce strip steaks and loin roasts.

I recommend 1 to 1¼ inches for loin steaks. Measure out, starting at the posterior end, making light marks with your knife at the corresponding cut points. Using a stiff-bladed knife, cross-cut steaks, one at a time, by cutting through the *longissimus dorsi* and the surrounding muscles, down to the bone. Repeat on the opposite side with the tenderloin. Complete by sawing through the finger and vertebrae bones using a handsaw. (If you have access to a bandsaw, you can use it to cut steaks to a uniform thickness. Start from the posterior end and scrape the bone dust afterward.)

INCLUDED CUTS: porterhouse steak, T-bone steak, club steaks.

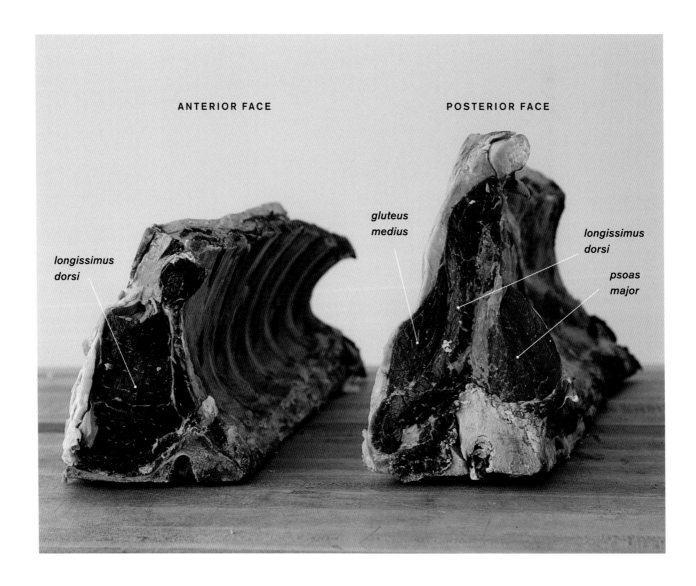

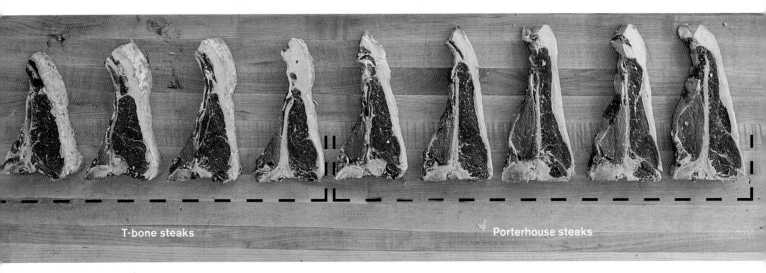

T-bone steaks | Porterhouse steaks

T-BONE STEAKS

T-bone steaks are cut from the anterior half of the short loin. The muscles included in a T-bone are identical to those in the porterhouse, but they are smaller due to the tapering of both the *longissimus dorsi* and the tenderloin. T-bone steaks have a minimum width requirement of ½ inch, which is about the diameter of a U.S. dime. Steaks cut from the remaining portion of the short loin, with little to no remnant of the tenderloin, are considered club steaks, which are more or less bone-in strip loin steaks.

PORTERHOUSE STEAKS

Porterhouse steaks are bone-in steaks that include both the dorsal muscles of the short loin – *longissimus dorsi* and *multifidus dorsi* – and the muscles of the tenderloin – *psoas major* and *psoas minor* – separated by the finger bones of the vertebrae. The tenderloin tapers along the short loin from the sirloin end (posterior) to the rib end (anterior). The cross-section of tenderloin on a porterhouse steak must have a minimum width of 1¼ inches, which is a little wider than the diameter of a U.S. quarter. Therefore, porterhouse steaks come from the posterior half of the short loin, where the tenderloin is wider. The last porterhouse, which is the one cut from the very posterior end of the subprimal, is referred to as a "vein steak," for the presence of the thick connective tissue between the *longissimus dorsi* and *gluteus medius*.

Strip Loin Roast, Bone-In

ALTERNATE NAMES: top loin roast, loin roast

Bone-in roasts are cut exactly the same way that a bone-in steak is cut, just thicker. The roast can be denuded or roasted with the fat cap in place. For a rough judge in portioning, assume that each finger bone can feed two people. After you decide on your roast length, then use the remainder of your strip loin to prepare strip steaks, either bone-in or boneless.

The strip loin is the remainder of the short loin after the tenderloin is removed and the protruding edge of the vertebrae is trimmed, leaving mainly the *longissimus dorsi* and *multifidus dorsi*. The strip loin cut is the starting preparation used for loin roasts as well as for bone-in and boneless strip steaks.

Before you can prepare a bone-in strip roast or cut bone-in strip steaks, you need to remove the section of the vertebrae that was connected to the tenderloin. Lay your short loin on the sirloin (posterior) end. Remove the protruding edge of the vertebrae by sawing between it and the flat of the finger bones.

Strip Loin Roast, Boneless

ALTERNATE NAMES: boneless top loin roast, boneless loin roast, strip filet, strip petite roast

Boning the strip loin will give you the option of making boneless roasts or steaks. Although you can remove the edge of the vertebrae, this is an unnecessary step if you plan on boning this whole subprimal. After following these instructions for boning and denuding the strip, you can prepare boneless roasts. For smaller roasts, you have two options: you can cross-cut the strip into smaller sections, or you can separate it horizontally into two thinner, long roasts. The latter option, which could also be cross-cut further into smaller roasts, should then be tied to maintain its size and shape.

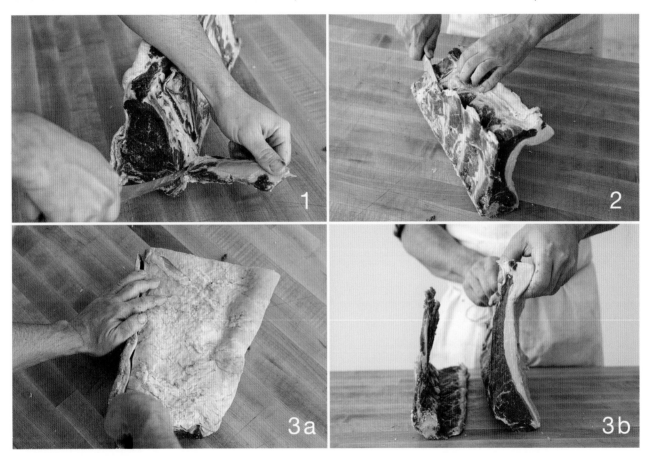

1. REMOVE THE THIRTEENTH RIB.
Get your blade underneath the rib and follow the flat face of the bone to its connection with the vertebrae. Either cut the rib out of its socket or twist it with your hand and tear it away.

2. FOLLOW THE FINGER BONES.
The tips of the finger bones are soft cartilage, so make sure you keep those with the bone. The bones also change in height as you move toward the rear of the strip, so it is helpful to start with the longest one. First, cut along the tips of all the bones, pulling the meat away from the bones. Then follow the flat face of the bones to their connection with the chine bone. Navigate your knife over the ridge at the base of the bones and stop when you reach the feather bones.

3. CUT ALONG THE FEATHER BONES.
Place the strip fat-side up. Find the tip of the feather bones and cut along it, staying against the bones. Connect this incision to the one made along the finger bones. Doing so will finish the removal of the boneless strip. Trim away any connective tissue or bone sheath that may remain from the finger bones. Trim the exterior fat to a desired thickness.

LOIN

Denuding the Strip Loin

The strip loin can be trimmed of fat and sinew before cooking, or you can leave it on. This trim work can be done to a whole strip loin or individually with portioned cuts. The loin muscle has a thick layer of silverskin that runs along the outer (dorsal) face and beneath the overlying fat cover. The strip loin is thicker on the bottom (medial) edge and thins out as it goes across. On the outer face of the muscle, the grain runs from the anterior to posterior end; try to work in this direction when removing the silverskin. Also, try to keep the silverskin in one piece, which allows for a much cleaner removal. The strip loin presents one of the more challenging denuding tasks, so tailor your expectations accordingly.

1. PEEL AWAY THE FAT.

The silverskin sits underneath the layer of fat, which can be peeled away by hand. Get underneath the fat layer and pull it away while keeping the muscle pinned to the table. Use your knife with any stubborn areas.

2. SEPARATE THE THICKER EDGE.

Start at the bottom edge of the silverskin, where it is thickest. Work your knife underneath the edge of the silverskin, down the length of the muscle.

ADDITIONAL PROCESSING

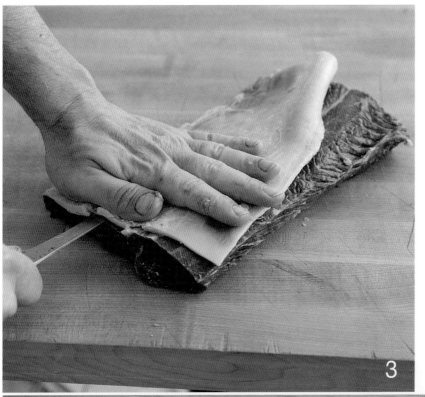

3. WORK ACROSS THE MUSCLE.

After you have separated the thicker edge, work across the muscle, following the curvature of the loin muscle while keeping your blade slightly angled toward the silverskin. Keep going until the silverskin is fully removed. At the middle of the muscle there is a ridge of silverskin that goes into the muscle. Cut across it rather than trying to remove it. At the posterior end you will hit a thick piece of silverskin that runs underneath a small portion of the *gluteus medius* muscle, which lies atop the *longissimus dorsi*. You may choose to remove this small muscle remnant and the underlying silverskin, or leave it. When it's left in, steaks from this area are known as "vein steaks" due to the presence of the silverskin.

DENUDING THE STRIP LOIN

Strip Steaks, Bone-In

ALTERNATE NAMES: New York steak, Kansas City steak, strip loin steak, shell steak, club steak

Strip steaks are cross-cut steaks from the strip loin and contain the *longissimus dorsi* and *multifidus dorsi* muscles. Center-cut strip steaks are prepared from the portion of the strip loin anterior to the *gluteus medius* muscle, thus avoiding the inclusion of the thick connective tissue underlying that muscle. Steaks that include the *gluteus medius* are considered vein steaks.

1. TRIM THE TAIL.
Start with a strip that has been trimmed of the excess vertebrae. Trim the top of the strip a few inches from the loin muscle, creating a tail on the steak that is an inch or two in length.

2. MARK YOUR STEAK.
Decide on your steak thickness (I prefer at least 1 inch), and then measure it out from the end of the strip. Cut through the muscles, down to the bone.

3. SAW AND REPEAT.
Once the muscles are completely severed, saw through the bones. Avoid contact with the meat to prevent shredding.

Strip Steaks, Boneless

ALTERNATE NAMES: boneless New York steak, boneless Kansas City steak, boneless strip loin steak, boneless shell steak, boneless club steak

Boneless strip steaks are easy to cut from any boned strip. Start by following the preceding instructions for boning the strip. Place the boneless strip on its fat cap. Decide on a steak thickness and measure out the first steak, starting from the anterior end. Place your hand against the flat face and, using a long-bladed knife, cut the steak in one motion to avoid any serrations on the steak face. If you don't want to bone the entire strip, or if you just want a couple of boneless steaks, follow the instructions for cutting bone-in steaks (see page 268) and then remove the bone afterward.

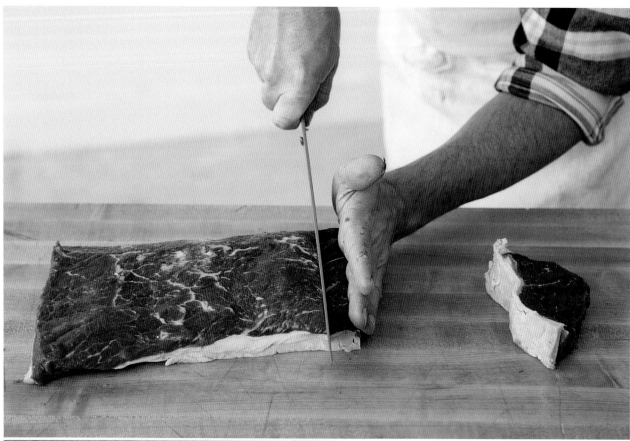

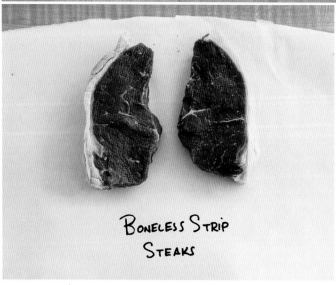

Boneless Strip Steaks

Tenderloin

ALTERNATE NAMES: short tenderloin, filet mignon, tournedos, chateaubriand, tenderloin medallions

The tenderloin gets its name from its defining characteristic: tenderness. The three main muscles that make up the tenderloin – *psoas major*, *psoas minor*, and *iliacus* – are the muscles used most infrequently by the animal. These muscles sit underneath the lumbar vertebrae, and originate from the pelvic girdle. In humans these muscles help maintain an erect posture. By contrast, in quadrupeds these muscles are rarely needed and, hence, rarely exercised. This results in a composition of very fine muscle fibers with little to no interior collagen or fat. While the tenderloin is prized for its texture, its flavor is lean and elegant; yet it lacks some of the bolder, beefier flavors of more used muscles.

The following instructions provide you methods for utilizing the full tenderloin. However, these instructions can be easily adapted for working with a short tenderloin (the portion removed from a short loin).

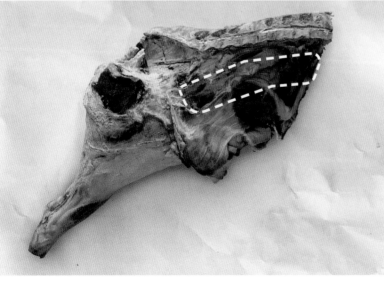

TENDERLOIN

Denuding the Tenderloin

Traditionally the tenderloin is prepared in two ways — as portioned steaks or as a roast. But before either preparation can be made, this muscle grouping must be cleaned. Working with the tenderloin is unique: it is a delicate grouping, and you must take care not to damage the fine muscle fibers. Proceed through these steps with that caution in mind.

1. PEEL AWAY THE MEMBRANE.
The first step is to remove the exterior fat, membranes, and *psoas minor*, which is also called the chain of the tenderloin. Most of this first step can be performed without knife use, but be careful not to tear the delicate meat. Get your fingers underneath the enveloping membrane and work it off the muscle; cut it away at the butt end.

2. REMOVE THE CHAIN.
Separate the chain (*psoas minor*), starting with your fingers and then finishing with your knife. Save the chain for grind; as an alternative, denude and pound it flat, and then cook it as a thin steak similar to flap meat.

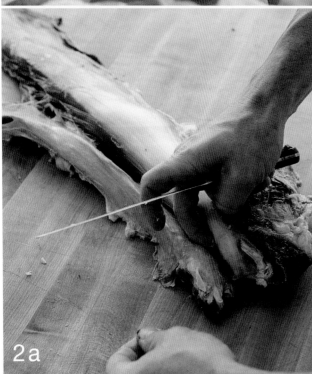

ADDITIONAL PROCESSING

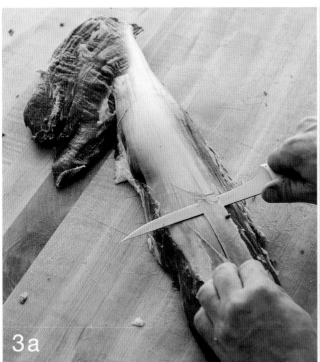

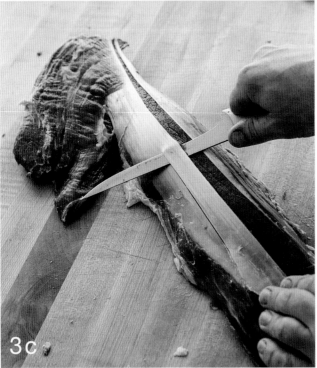

3. DENUDE, IN SECTIONS.

Get your knife; a thin fillet knife works great for this procedure. Whatever knife you use, make sure that it's sharp. Slip the blade underneath a strip of silverskin, starting at the anterior point. It's helpful to place your free hand on the tenderloin, behind your knife of course, to help stabilize it. Angle your blade slightly toward the silverskin and cut all the way out the butt end. Once removed on the butt-end, circle back and cut the strip off at the other end. Repeat this process, removing sections of the silverskin, until it's removed. You will notice that the silverskin slopes between the area where the foot connects. Carefully remove this area of the silverskin, cutting between the two muscles but not so much that the foot is separated.

TENDERLOIN

Denuding the Tenderloin CONTINUED

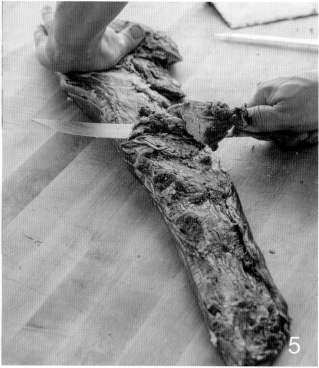

4. CAREFULLY CLEAN THE FOOT.
The silverskin slopes between the *psoas major* and *iliacus* at the sirloin butt end. Carefully work the tip of your knife underneath the silverskin and follow the curvature between the two muscles, finishing when the silverskin dissipates and before you completely separate the two muscles.

5. CLEAN UP THE UNDERSIDE.
You will find bits of connective tissue and rough fat where the tenderloin attached to the finger bones. Traditionally the underside is cleaned up for a fancier appearance, but you can choose to leave it on, as the fat will help add some flavor. To clean, flip the tenderloin over and work with the grain, starting at the butt end. Cut away as little as possible until you have a consistent face.

TENDERLOIN

Tying the Tenderloin

The tapered shape of the tenderloin provides a slight challenge for roasting it whole: the anterior tip will be drastically overcooked when the sirloin butt end is properly done. This problem is easily remedied by tying your roast. Find a point, about 5 inches from the tail end, that, when folded over, will help achieve a more even thickness for the entire piece. Then cut about 80 percent through at this point, which then allows you to fold the tail under. The result will be a flat, clean face and an even thickness. Tie the roast in two passes: the first pass at every 3 inches and the second pass in between them, starting at the tail end. Be careful not to overtighten your knots, as the string can cut into the meat due to the delicate nature of the muscle.

TENDERLOIN

Portioning Tenderloin Steaks

The filet mignon may be the most well-known cut from a beef carcass. The extreme leanness of the cut lends itself well to barding, the common preparation of wrapping with bacon. This boneless cut is easy to portion.

Clean, denude, and tie a tenderloin per the aforementioned instructions. Tie the tenderloin in increments that correlate to the desired steak thickness. Cross-cut into steaks, using a long-bladed knife, in between every one or two ties. Place one hand against the meat when cutting to ensure a straight and even cut. Cut steaks no less than ¾ inch thick; cut medallions about ½ inch thick.

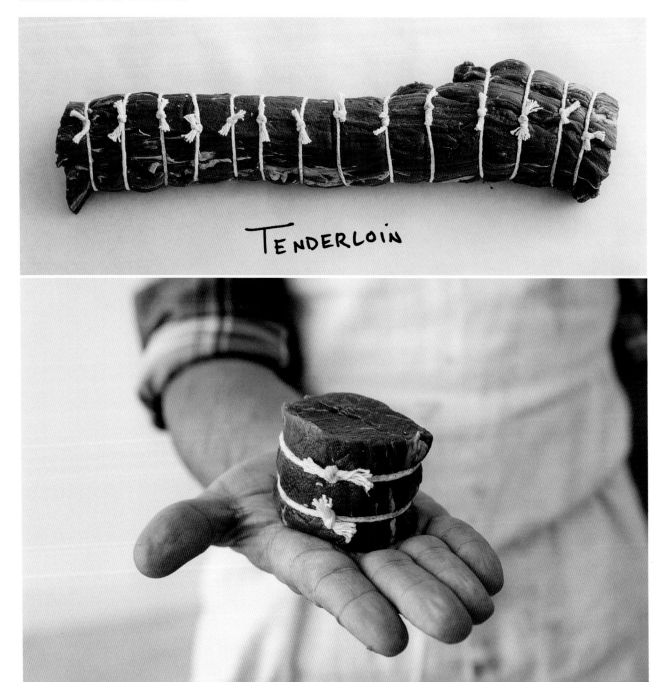

Top Sirloin Cap *biceps femoris*

ALTERNATE NAMES: coulotte, sirloin strip, sirloin top butt steak, rump cover, rump cap, picanha, sirloin cap

The sirloin cap is part of the *biceps femoris* muscle, known also as the outside round. While said muscle is characteristically tough in the round primal, this dorsal portion of the muscle tends to be far more tender, lending itself to other culinary preparations, like the classic skewered-and-grilled approach in Brazilian cuisine.

The sirloin cap needs some minor cleanup before cooking. Denude the meat side that was attached to the top sirloin center. Follow that by trimming the fat cap to your desired thickness, or removing it completely if you wish.

Sirloin Cap (Coulotte)

Top Sirloin Center *gluteus medius*

ALTERNATE NAMES: top sirloin butt center, top sirloin steak, sirloin steak, center cut sirloin steak, filet of sirloin, baseball steak

Instructions for removing the accessory muscles of the top sirloin butt can be found on page 203.

The top sirloin center muscle, *gluteus medius*, is prized for its blend of tenderness and rich flavor, despite being a large, weight-bearing muscle. Furthermore, it's quite lean, making it an ideal candidate for ground beef that needs a lower fat content yet bold flavor. As an intact muscle, it can be processed in few different manners: roasted whole, cross-cut into center cut steaks, or seamed into two uneven sides, the smaller of which is known as the baseball.

Isolate the right two muscles, the smaller of which is the baseball cut.

SEPARATING THE BASEBALL

Seaming the top sirloin center involves following a meandering band of connective tissue, something that will certainly take some practice. The connective tissue spreads out in multiple directions; the thickest part is removed, leaving a thinner portion inside the muscle. After the muscle is seamed, the smaller section, coming off the dorsal side, is known as the baseball cut.

1. FIND THE BAND OF CONNECTIVE TISSUE.

Follow the wavering face down the length of the muscle, separating the baseball cut.

2. REMOVE THE COLLAGEN SEAM.

After the two sides have been separated, remove the thick band of connective tissue from the larger side; leave the thinner portion that juts off and is embedded in the muscle. Denude any remaining connective tissue from both sides of the muscle. Each portion can be roasted whole or cross-cut into steaks. The larger portion can be cut lengthwise into smaller roasts, or those smaller roasts can be cross-cut into medallions. Lastly, sirloin kabobs can be cut.

TOP SIRLOIN CENTER

Tri-Tip

ALTERNATE NAMES: Santa Maria steak, Newport steak

The tri-tip comes from a muscle, *tensor fasciae latae*, that spans across the juncture between the sirloin and the round; specifically, the tri-tip comes mainly from the bottom sirloin butt, though a portion of it will also be found atop the sirloin tip. It's a working muscle, helping with stability of the leg and operation of the patella.

The tri-tip needs minimal cleaning: denude the underside and trim the fat cap to a desired thickness. If the tri-tip is large enough, cross-cut steaks can be portioned from it. Find the grain direction and cut across the whole muscle, aiming for a thickness of at least 1 inch.

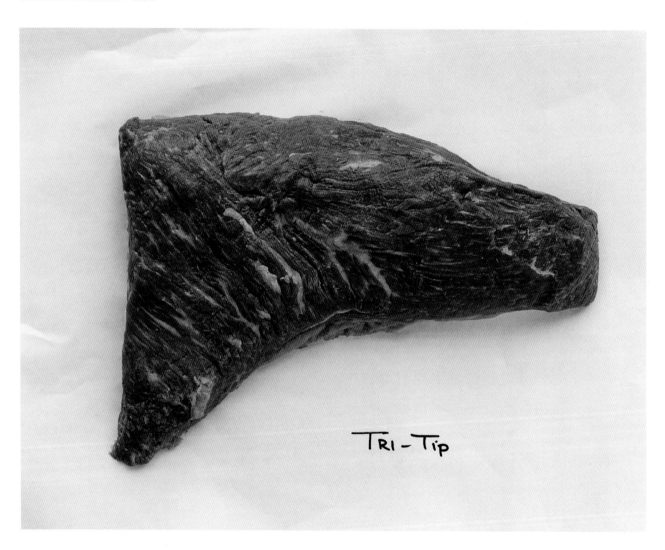

FLANK

THE FLANK PRIMAL (not to be confused with the flank steak within) is the only boneless primal. There are three cuts that come from this primal: the flank steak, the inside skirt, and the sirloin flap. As a group, they are sometimes referred to as "flap meat." They are all thin, strong muscles, doing the unrelenting work of holding up the abdominal viscera. They can benefit from some tenderization but, above all else, make sure to slice these thin and across the grain (see page 64).

INCLUDED CUTS: flank steak, inside skirt (see page 241), sirloin flap

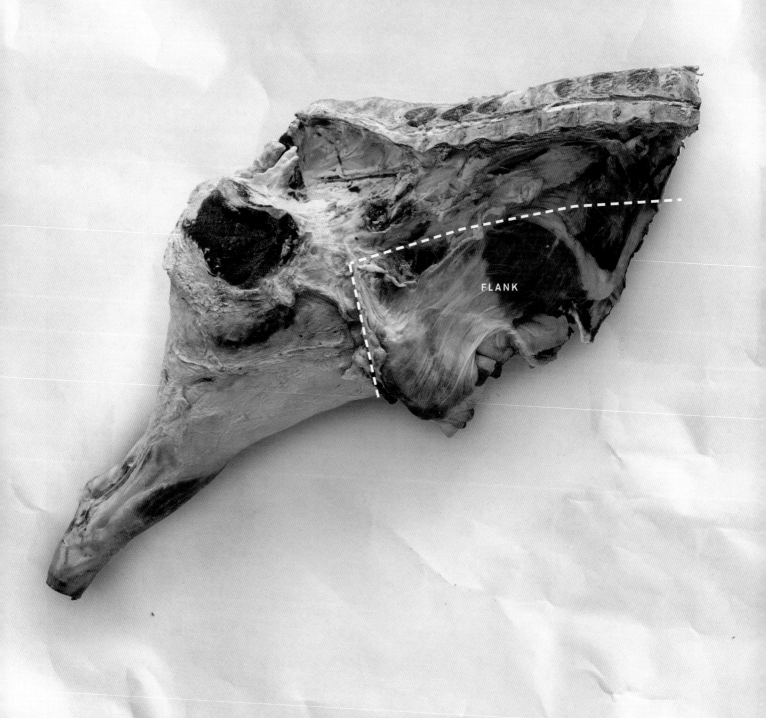

Flank Steak *rectus abdominis*

ALTERNATE NAMES: flap meat, jiffy steak, London broil

The flank steak is an easy, quick-grilling steak that, like all flap meats, benefits from tenderization and slicing on a bias across the grain. It was popularized as the original muscle for London broil, though that term has now become a moniker for many cuts.

CLEANING THE FLANK

The flank steak needs very minor prep work. At the thin end of the muscle there is a thick oval layer of silverskin. Use a knife to get under the silverskin at the edge of the muscle, and then remove the exposed area without cutting too deep. Square up any tapered edges that have connective tissue, most likely at the wider end. Last, score the surface on both sides with a shallow crosshatch pattern to prevent the surface tissue from curling while cooking.

FLANK STEAK

ADDITIONAL PROCESSING

Sirloin Flap *obliquus abdominis internus*

ALTERNATE NAME: flap meat, sirloin bavette

CLEANING THE SIRLOIN FLAP

The sirloin flap has a thin layer of silverskin on both sides, one on the dorsal edge and one on the ventral edge. The flap is a long, flat muscle, and so I recommend using a long-bladed knife to remove this silverskin. By starting with the base of the blade, nearest to the handle, you can use long strokes to slice underneath the silverskin, keeping the silverskin as a whole sheet. Angle the blade slightly at the connective tissue as you cut. As you release the silverskin, use one hand to keep the silverskin taut, which helps you to cut against it. Repeat the long strokes until the entire silverskin is removed, and then repeat on the opposite side of the muscle.

1. RELEASE THE TOP EDGE.

The best approach is to release one entire edge of the silverskin and then slice the whole sheet away with long strokes. Using a long-bladed knife, get the edge of the blade underneath a portion of the silverskin by laying the knife flat against the meat and pressing the edge of the blade slightly down into the meat.

Sirloin Flap CONTINUED

CLEANING THE SIRLOIN FLAP CONTINUED

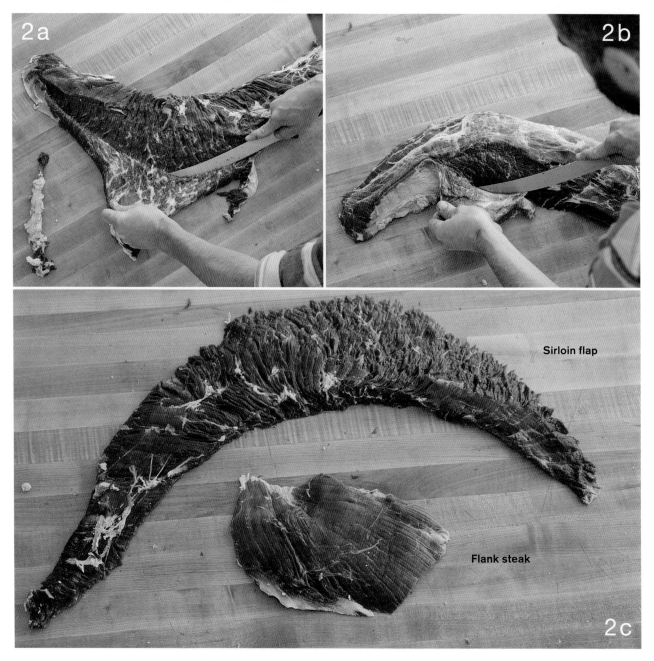

2. CUT THE SHEET AWAY.
Hold on to the released top edge of silverskin and continue to cut the sheet away. Keep the long blade flat against the meat. Repeat on the opposite side.

ROUND

THE ROUND: LAST AND, SORT OF, LEAST. It's by far the least interesting primal to cook from, producing lean, variably tender meat with moderate amounts of flavor. The muscles of the hind leg (which the round consists of) are some of the largest in the entire carcass, making them good candidates for portioning. Each of its subprimals, the bottom round, sirloin tip, and top round, require a fair amount of processing and are good to practice techniques on since their large size is fairly lenient, meaning you won't lose the whole cut to mistakes.

INCLUDED CUTS: eye of round, heel, outside round, outside round steaks (western tip, western griller), sirloin tip, shank, top round, top round steaks (San Antonio, Santa Fe, Tucson)

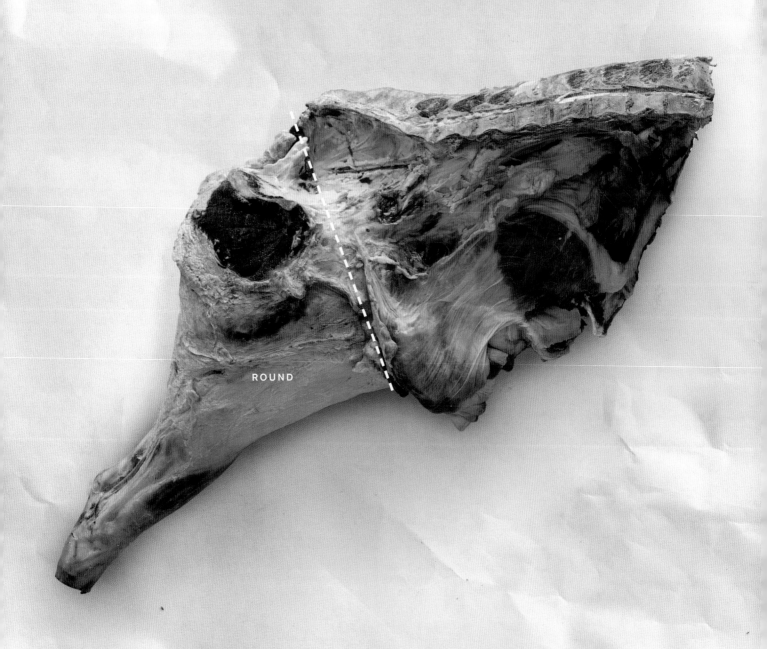

ROUND

Eye of Round *semitendinosus*

The eye of round is a lean, oblong muscle that slightly resembles the shape of a tenderloin, though its texture is drastically tougher. Its best uses are for thin-slicing roast beef or cured for bresaola.

CLEANING: The eye of round has minimal connective tissue on its surface and, depending on your cut, can be prepared almost as is. For a leaner, cleaner preparation follow the basic guidelines for trimming connective tissue (see page 54) and tidy up to your desired leanness.

Heel

ALTERNATE NAMES: merlot cut, braison cut, Pikes Peak roast, horseshoe roast

The heel is the toughest cut to come out of the bottom round subprimal, as it sits closest to the ground and, therefore, where the animal worked its hardest. It is composed of two muscles, the *gastrocnemius* and the *superficial digital flexor*, the former being the larger of the two. The composition of the heel is similar to shank meat, and it is frequently trimmed and added to ground round. But with some seaming and trimming you can get some unique preparations out of this sinuous muscle.

Seaming the Heel

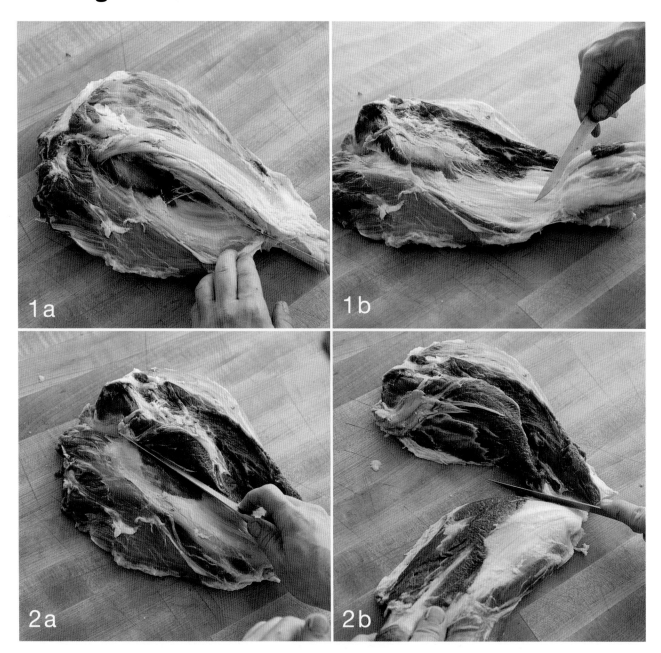

1. REMOVE THE FLEXOR MUSCLE.
Locate the *superficial digital flexor* and remove it through natural seams. It has a strong covering of connective tissue that can be trimmed away. You can either cross-cut this muscle for braising-bound portions or add it to the grind pile.

2. SPLIT THE *GASTROCNEMIUS*.
The remaining *gastrocnemius* can be further separated into two segments, medial and lateral, based on another natural seam found beneath the former. A sheath of connective tissue best identifies the medial section. Find the seam running beneath the *gastrocnemius* and separate the two segments.

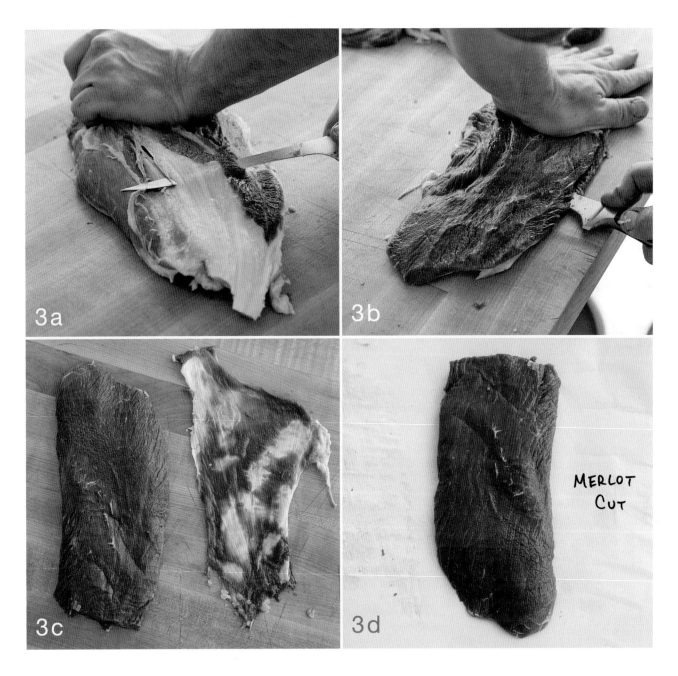

3. DENUDE THE MEDIAL PORTION.

The medial section can be further cleaned of exterior connective tissue, resulting in the merlot cut. The lateral section is best added to grind or stew. Start by cleaning the interior face of the muscle, getting underneath the silverskin and removing it. Remove the exterior silverskin by laying the muscle flat and running your knife underneath it, staying parallel to the surface.

Outside Round *biceps femoris*

ALTERNATE NAMES: silverside, gooseneck round, bottom round, outside round (flat)

The outside round sits on the outside of the hind leg. This is a working muscle, giving it a tougher texture with more connective tissue than the other round subprimals that surround it. Therefore, the outside round needs to be tenderized, mechanically or chemically. Its lean composition makes it a frequent choice for stew meat or kabobs; thin steak preparations, such as chicken fried steaks; and other cuts that benefit from minimal interior marbling. Whole, it is often used for either a pot roast or long-term transformations, such as curing, pickling, or corning. Its leanness also makes it an excellent choice for beef jerky and other dehydration products.

ADDITIONAL PROCESSING

OUTSIDE ROUND

Denuding the Outside Round

The outside round has two main sheaths of thick connective tissue that may need to be removed, depending on your preparation. The first is what gives this cut its "silverside" moniker, as it runs almost the full length of the ventral side.

1. REMOVE THE SILVERSKIN.
Using a long-bladed knife, work to separate the silverskin from the muscle. Start at the upper edge and work down toward the table in long strokes, keeping the blade edge angled slightly toward the connective tissue. Repeat until the silverskin is removed in one piece.

2. SEPARATE THE HEADS.
The second strip of connective tissue separates the two parts of the *biceps femoris*: the main muscle (also known as the long head) and the flat triangle (also known as the ischiatic head). Removal is optional but is advised if you plan on slicing thin steaks or preparing any quicker cooking cuts. Find the band of connective tissue and separate the side muscle through the seam, using long strokes and keeping your blade edge angled slightly toward the silverskin.

OUTSIDE ROUND

Denuding the Outside Round CONTINUED

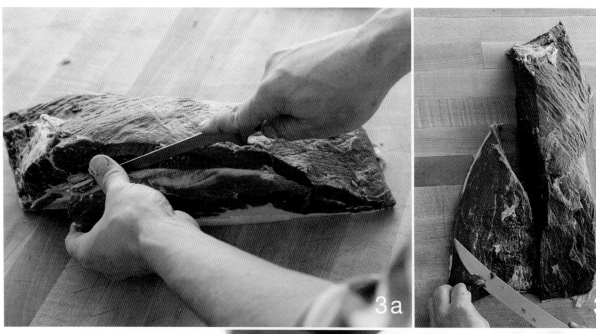

3. REMOVE THE SILVERSKIN.

Start at the exposed edge and work your way down the sheet of connective tissue. If you prefer, the surface fat and the underlying connective tissue can be trimmed away, which makes for leaner preparations. Lay the outside round on the table and run your knife underneath it, between the muscle and connective tissue, keeping your knife parallel to the table surface.

WESTERN TIP AND WESTERN GRILLER STEAKS

THE LONG HEAD of the outside round can be further portioned based on changes in muscle structure that occur from posterior to anterior ends.

THE TOP (PROXIMAL) END, nearest to the sirloin, can be separated and used as a rump roast or cross-cut into western tip steaks.

THE CENTER PORTION can be used as a roast or cross-cut into thin steaks, known as western griller steaks.

THE BOTTOM (POSTERIOR) TIP is the toughest and can be trimmed, squaring up the end, and used for stew or grind.

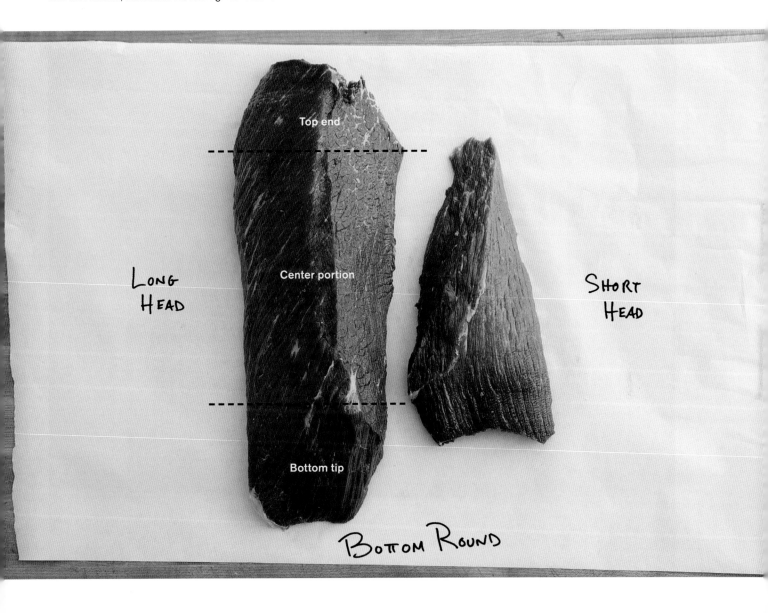

Sirloin Tip

ALTERNATE NAMES: knuckle, bottom sirloin, bald tip, ball tip, round tip, sirloin butt, tip roast, minute steak, sandwich steak

The general tenderness of the four muscles that make up the sirloin tip allow for versatile preparations. The sirloin tip contains the most tender muscle in the round — the *rectus femoris* — along with the *vastus lateralis*, the *vastus intermedius*, and the *vastus medialis*, the latter being the most tender of these three. (These four muscles are the bovine equivalent of the quadriceps.) There is also a portion of the tri-tip (*tensor fasciae latae*) that lies atop the sirloin. The sirloin tip will come in different sizes, depending on where you chose to separate the round and sirloin (see page 147), which, in turn, also affects the size of the ball tip in the top sirloin butt (see page 202). Notwithstanding, the following directions will apply no matter the size of the sirloin tip.

SIRLOIN TIP

Seaming the Sirloin Tip

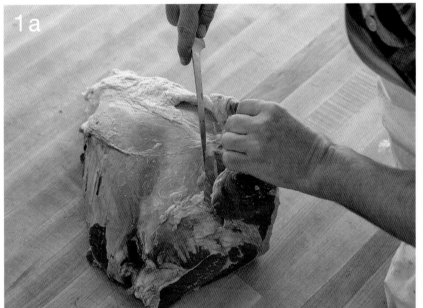

1. REMOVE THE TRI-TIP.
Start by trimming away the remnants of the tri-tip and exterior fat. If the tri-tip is large enough, save for a small steak; otherwise, grind it.

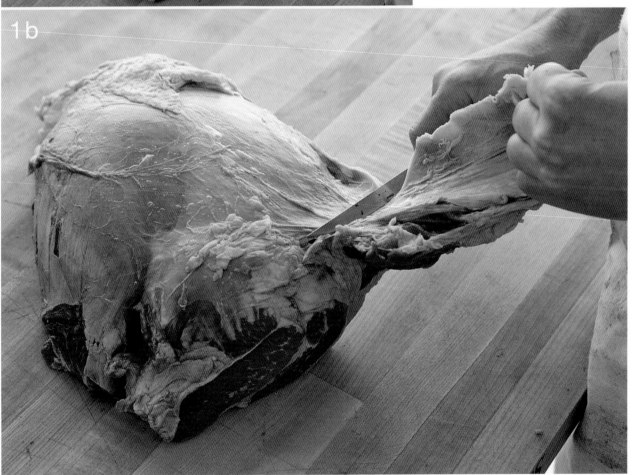

SIRLOIN TIP

Seaming the Sirloin Tip CONTINUED

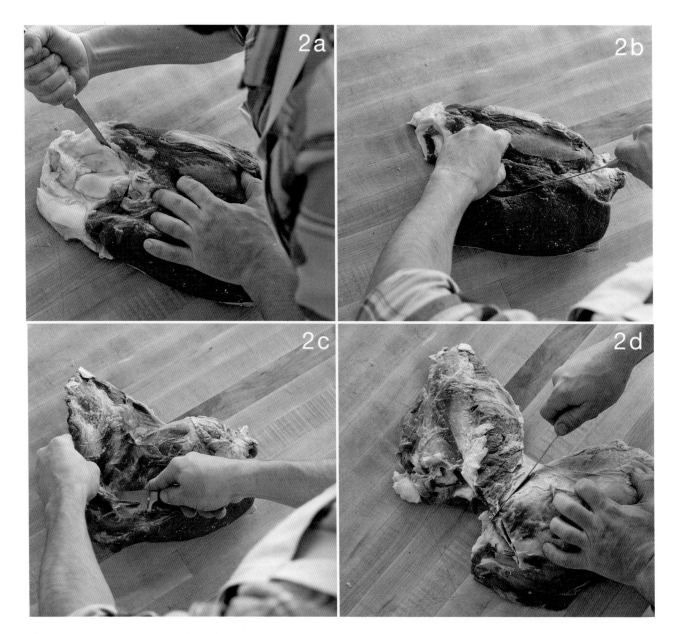

2. REMOVE THE INTERIOR MUSCLES.

Turn the sirloin tip over and if the patella is still attached, cut that away. Trim away the two smaller muscles that sat next to the femur, the *vastus medialis* and *vastus intermedius* muscles, following their natural seams. The *vastus medialis*, the larger of the two, can be trimmed of any exterior connective tissue for use as a quick-searing steak or kabob meat. The *vastus intermedius* should be cleaned and saved for stew meat or grind.

The two main muscles – the *rectus femoris* and the *vastus lateralis* – of the sirloin tip remain connected and are called the center roast and side roast, respectively. They can be roasted together, cross-cut into thin or thick steaks, or seamed and treated individually. The seam that separates the two muscles is pretty obvious, though it will appear that there are three seams due to the connective tissue present in the middle of the *rectus femoris* muscle. It is a tough seam to follow accurately due to the curvature of the *rectus femoris*.

3. SEPARATE THE MAIN MUSCLES.

After identifying the correct seam, start on the interior and follow it along the curvature of the *rectus femoris*. Once separated, continue cleaning both muscles. The *rectus femoris* can be butterflied to remove the medial piece of connective tissue. Both the *rectus femoris* and *vastus lateralis* can be roasted whole or cross-cut and tenderized for portioned steaks.

SHANK

THE SHANKS, HIND AND FORE, are a collection of muscles that, due to their proximity to the ground and the animal's constant working of them, contain more connective tissue than any other area of the carcass. They make the ultimate slow-and-low braising items, because during cooking the breakdown of copious amounts of collagen produces gelatin. This gelatin makes for an unctuous sauce and a self-basting environment, which keeps the meat moist and palatable.

CLEANING: Shanks are the areas most susceptible to contamination during slaughter and will often need to be cleaned of areas that were missed during post-slaughter inspections. Go over the shanks closely, looking for and removing hairs, manure, and other inedible remnants.

Boning the Shanks

Shanks can be boned for a boneless roast or, more likely, ground beef. The high levels of collagen make ground beef from shanks perfect for stewed preparations, like chili.

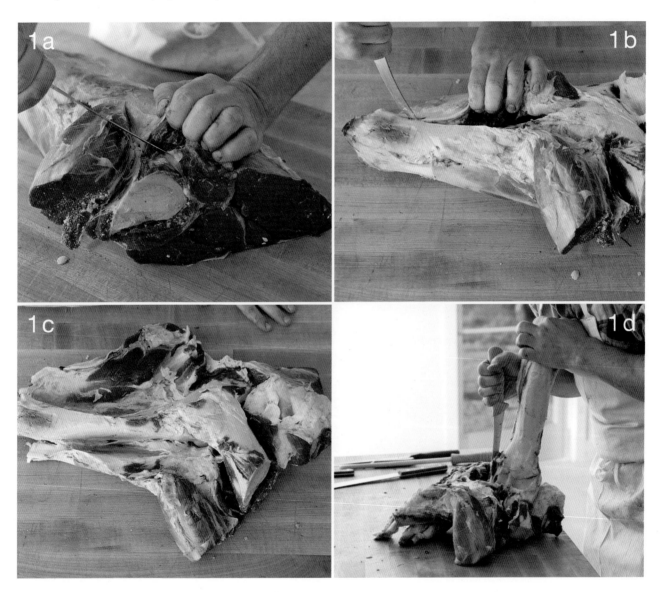

Follow the bone, top to bottom. With the foreshank, as shown in the image, start at the top (dorsal) where it was removed from the humerus. Work down both sides of the bone before peeling the remaining meat from the bone. (Shank muscles are hardy and peel away from the bone easily once areas of connective tissue are severed.)

When boning the hindshank, start at the area of bone that is exposed, working the flexible knife tip along the bone from top to bottom while peeling back the muscle. Continue around the leg bones until all the shank muscles are released in one piece.

SHANK

Making Cross-Cut Shanks

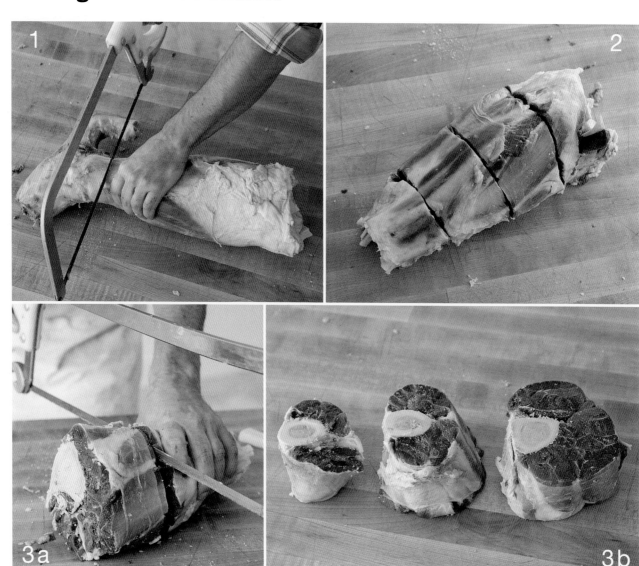

1. REMOVE THE GAMBREL.
The bottom (ventral) end of the shank needs to be removed, along with the attached gambrel cord. Saw through the bone where the meat ends to remove the gambrel.

2. MARK YOUR CUTS.
Place the shank bone-side down, and then measure out where you are going to cut the shanks. The best shanks come from the middle 60 percent or so of the shank. Cut through the muscle down to the bone.

3. SAW.
Roll the shank forward 90 degrees so that the bone is now facing you. Separate the sections of shank by sawing through the bone where you cut through the meat.

Top Round

ALTERNATE NAMES: inside round, topside, breakfast steak, butterball steak, London broil, minute steak, round steak, Santa Fe cut, beef round petite tender, San Antonio steak, Tucson cut

The top round subprimal provides the largest amount of tender beef in the round, primarily due to the inclusion of the *semimembranosus*, the largest muscle in the top round. (The muscles in the sirloin tip are certainly more tender, but they are much smaller in size.) Lean, tender, and versatile, qualities that accommodate almost any culinary method, the *semimembranosus* is the epitome of the round. The sheer size of the *semimembranosus* makes it a natural choice for roast beef; raw, it is the ideal choice for tartare; tenderized, its thick steaks pack more flavor than the renowned tenderloin.

The top round is made up of five muscles. From largest to smallest, they are the *semimembranosus*, the *gracilis*, the *adductor*, the *pectineus*, and the *sartorius*. These muscles can be seamed out and the first three used for their individual benefits. The smaller two are best saved for grind or stew meat.

CLEANING: The top round can be used for roasts and large top round steaks with minimal cleaning. Trim the top round fat cover to a desired thickness. On the internal surface trim away any inedible connective tissue, blood vessels, or fat deposits.

TOP ROUND

Seaming the Top Round

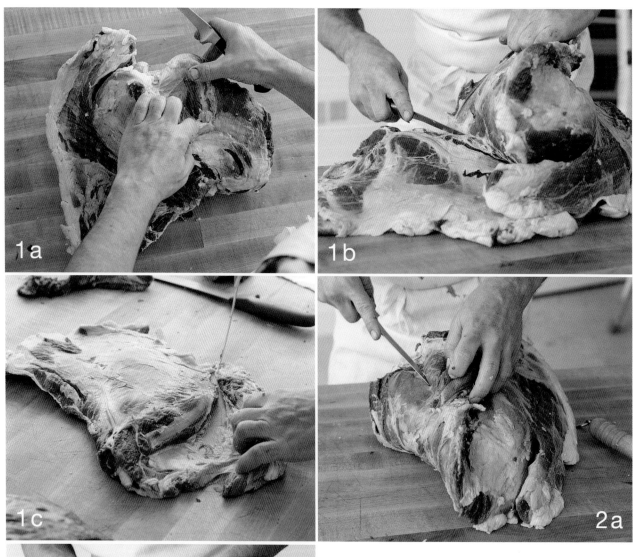

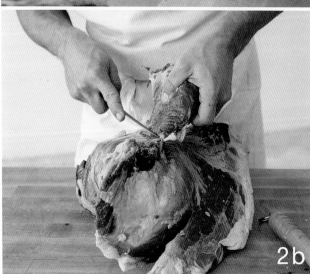

1. REMOVE THE CAP COMPLEX.

It's easiest to remove the top round cap (*gracilis*) with the *sartorius*, the smallest side muscle of the top round. Find the seam between the *sartorius* and *pectinius*. Follow that seam to the top round cap and then continue removing the two muscles together based on the seam beneath the top round cap. Then separate the two muscles.

2. REMOVE THE *PECTINIUS*.

The next small side muscle is the *pectinius*. Remove it through the seam adjacent to the *adductor*.

ADDITIONAL PROCESSING

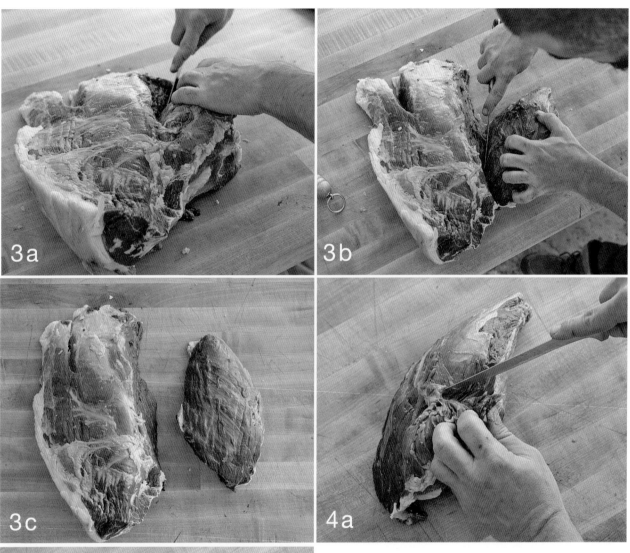

3. SEPARATE THE *ADDUCTOR*.

The remaining portion is considered a cap-off top round. It contains the *semimembranosus* and the *adductor* muscles. Roasts and steaks cut from the cap-off top round provide a cleaner presentation and result in a more palatable product than those that utilize the entire subprimal. The remaining seam is easy to access and follow, as the muscles are large and more loosely kept together. Separate the two muscles, accessing the natural seam from the internal surface.

4. REMOVE ARTERIES FROM THE *ADDUCTOR*.

The *adductor* has some prominent arteries that need to be removed. Locate the arteries and cut them out, removing as little meat as possible.

SEAMING THE TOP ROUND

ROUND

Seaming the Top Round CONTINUED

5. FINAL CLEANING.

Denude and trim up the five seamed muscles. Remove the silverskin from the sloping end of the *semimembranosus* as well as the fat cap. The *gracilis* can be denuded in the same way as the sirloin flap (see page 283), laying the blade flat against the meat and using long strokes to remove the entire sheet of connective tissue.

PORTIONING THE TOP ROUND

STEAKS: Steaks can be cut from a whole top round subprimal or individual muscles. Cross-cut at preferred thickness for steaks: ½ inch for quick-cooking minute steaks; ¾ to 1¼ inches for top round steaks; 2 inches for London broil. The *gracilis*, when left whole and cooked thin, is marketed as a Santa Fe cut. The *adductor,* when cut into steaks, is called the San Antonio steak, while the Tucson cut uses the *semimembranosus* (shown below). Palatability of all top round steaks is aided by some method of tenderization, ideally mechanical, because chemical tenderization may cause the lean, small-grained structure of these muscles to turn mealy.

ROASTS: The top round can be roasted whole or cut to any smaller-sized roast. Individual muscles can also be roasted whole or in portions, although the *semimembranosus* and the *adductor* are the best options for single muscle roasts. The *gracilis* can be rolled, tenderized, or even stuffed for a top round cap roast. All of these roasts benefit from tying, which helps maintain shape and promotes consistent cooking.

The Tucson cut

BONES

ALL BEEF BONES should be saved because, no matter their origin, they can serve some purpose. There are two main uses for bones — stock and marrow — with a third minor use being dog bones.

STOCK BONES: The best stock bones are those with copious amounts of collagen. Collagen is the main protein in connective tissues that, when hydrolyzed, converts to gelatin, one of the foundational elements of good stock and soup bases. Good stock bones include the vertebrae and joints. The ribs, scapula, and pelvic bones can be used for stock but will add little additional flavor. Stock bones need no additional cleaning and should be used as is, with scrap meat and connective tissue left attached.

Marrow Bones

Within the skeleton there are, generally speaking, flat bones and round bones (ribs and femurs are examples of each). The interior of round bones is filled with a fatty substance known as marrow that, when roasted, is delicious and spreadable. (The flat bones have a form of marrow in them as well, but it's not useful for culinary applications.) The marrow bones, in order of preference, are the femur, tibia, radius/ulna, and humerus. The marrow sits inside the bone in an area called the *medullary cavity*. To access the cylindrical medullary cavity, you must either cross-cut the bone into segments or split the bone in half lengthwise. The former is the more common method.

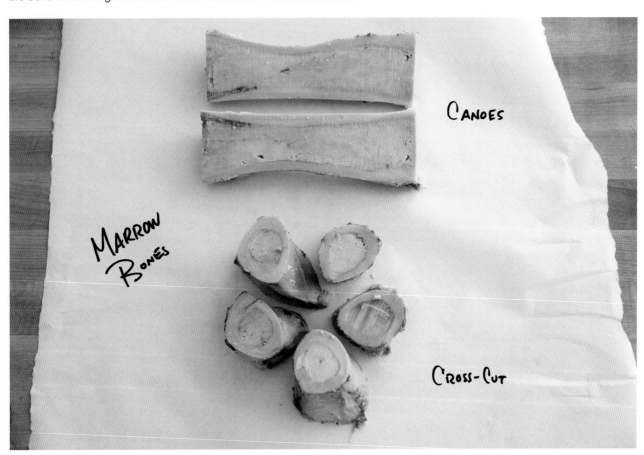

CUTTING: Remove the joint end of the bone, sawing where the shape of the joint transitions into the barrel of the round bone body. Cross-cut every 2 to 3 inches, choosing a measurement that maximizes yield. Lengthwise cutting is, aside from being dangerous, really only possible if you have access to a bandsaw. To make a lengthwise cut, remove both joint ends of the bone. Then secure the bone between your hands and carefully split the bone in two, being careful not to let the bone roll.

CLEANING: Marrow bones can be cleaned of the fascial covering. This cleaning is purely for presentation purposes, so feel free to skip this step if your preparation does not need it. All bones have a fascial covering, which is sometimes called bone sheath or bone tissue. You will have seen it come off when peeling muscles away from bones, such as the top blade or sirloin tip. Bones can be cleaned before or after cutting, though handling and cleaning a single large bone rather than individual small bone segments is certainly the easier method.

Use a boning hook to clean, as the sharp tip will get under the sheath and the dull body will help pry it away from the bone. Scrape down the bone repeatedly, using a knife to clean areas where the fascia has adhered to the bone. Do not use a knife that you are trying to keep sharp. Scrape until you have exposed the entire smooth bone surface.

VARIETY MEATS

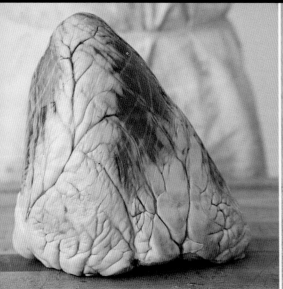

HEART

The cardiac muscle present in beef heart is the leanest muscle in the carcass. The heart can make for surprisingly delicious preparations and is versatile to boot. It may be considered a good gateway organ for offal-eating beginners: its taste is rich and flavorful with only undertones of organ, unlike the liver or kidney.

CLEANING: The heart's exterior is dappled with fat, while the interior chambers and vascular system are lined with a tough fascia. All of this needs to be removed. Start by separating the chambers based on thickness. This method allows you to work with flattened pieces and helps ensure that the final product will cook evenly. Lay a piece with the fat exterior facing up and trim the entire surface. Use shallow strokes to expose the underlying lean muscle. Save the fat for rendering or for the grind pile. Flip the piece over and do the same with the chamber walls and vessels. The interior surface is uneven; bend and manipulate the piece of heart to allow better access for trimming. Repeat with remaining pieces.

LIVER

The liver is the offal that has turned most folks away from eating organs. Difficult to prepare, slimy to touch, and strong in taste, the current American palate has rejected this organ largely due to the past generation's improper preparation. These days the liver is often an ingredient in natural or raw dog foods, but I encourage you to give liver a try so that you can make up your own mind.

CLEANING: The liver requires minimal cleaning. Remove any remnants of the vascular system and bile duct connections. Inspect the surface for lesions and other unpalatable markings. Cut into manageable pieces, find a good recipe, and enjoy.

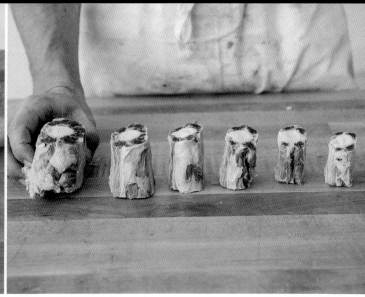

OXTAIL

Oxtail, while not an internal organ, is often considered a variety meat in America due to its unpopularity. It has a low meat-to-bone ratio, especially for pieces farther down the tail. But its high collagen content makes oxtail an ideal choice for stews and rich braises, or as a meaty addition to stocks.

CLEANING: Oxtail needs minimal cleaning. Carefully inspect it for areas of contamination that may have been overlooked during slaughter. Turn it over so that the underside of the tail is pointing up. Separate into pieces by locating the cartilaginous divisions between the bones, indicated by a soft, white marker. Separating bones should provide minimal resistance to the blade. However, if cutting presents a challenge, make adjustments to the angle of the blade or small changes in position.

TONGUE

The beef tongue is pure muscle and, considering that the animal constantly used it, this muscle has a deep flavor. The texture of the tongue tends to feel like a cross between brisket and something from the round. It is lean and has a nice grain. Upon experimentation you will find that it is delicious and versatile.

CLEANING: Thoroughly wash the tongue under cold water, scrubbing the papillae-laden surface with a brush. Remove any potential remnants of the rumen's contents. The exterior skin of the tongue needs to be peeled away, and this is easiest done after a long period of braising. Refer to a recipe for further tongue preparation.

SUET AND OTHER FATS

Suet is the hard, often crumbly, interior cavity fat that sits around the kidneys and loins. Its only raw usage is as feed for birds, though it is often rendered and rolled with seeds and other avian-adored ingredients. For any culinary usage it needs to be rendered, a process that converts the raw fats into edible tallow. Tallow production is an easy process that just takes time and a source of low heat.

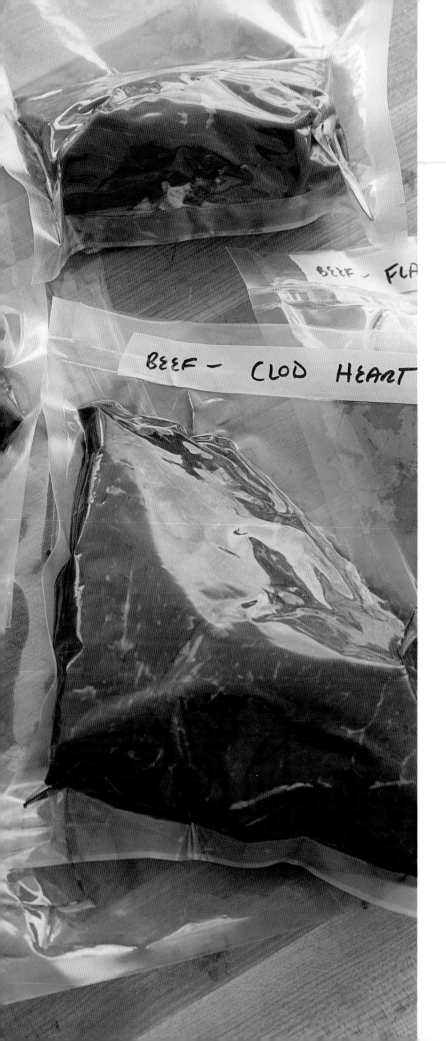

PACKAGING & FREEZING

PROPERLY PACKAGING AND STORING meat is essential for its preservation, enabling a harvest to be used over months or years. Failure to properly wrap and freeze meat products can lead to nutrient and flavor loss, absorption of foreign flavors, dehydration, and other textural changes. The irregular shape of meat products presents challenges when packaging. This chapter covers the main options for overcoming obstacles in packaging and freezing meat.

Freezing

FREEZING FOOD AS A FORM OF PRESERVATION has been around since prehistory, but for many areas of the world freezing was never an option until the advent of mechanical freezers. (Instead, salt was used extensively.) Today, freezing is the main method of meat preservation. Most home refrigerators come stocked with freezers that provide adequate functionality for short-term storage of meats. But for people looking to store large quantities for long periods, investing in a stand-alone freezer is the most effective solution.

The Effects of Freezing

As soon as foods are exposed to temperatures of 30°F or lower, their moisture begins to change from liquid to solid. This process aids in food preservation because at such cool temperatures the chemical and biological activities that cause spoilage slow down (though they do not cease) and liquid water is now trapped in a solid form, halting the microbes — bacteria, yeasts, and molds — that require water to continue their normal functions and propagation. This also means that the enzymatic activity that causes maturation and decay of the meat is slowed to an inconsequential pace. This lack of enzymatic activity is why freezing meat before resolution of rigor mortis will yield a tough, dry result. (For more on rigor mortis and cold shortening, see pages 16 and 17.) If conditions are maintained in perpetuity, these effects — halting the activity of microbes and enzymes — maintain the safety of the meat product, though extended periods of freezing will have adverse effects on quality.

Anywhere from 66 to 73 percent of the total mass of meat is water. This means that freezing causes the main component of meat to change from a liquid to a solid state; this shift has the potential to cause considerable damage. There are two types of water within meat:

- **EXTRACELLULAR MOISTURE** is water that exists outside of the cells. It is unadulterated and has few dissolved nutrients or other substances, such as salt, sugar, or minerals.
- **INTRACELLULAR MOISTURE** exists inside cells and contains all the dissolved components required for cells to operate.

When cellular components and other dissolved substances are in water, they crowd the area that water molecules have, getting in the way of bonds forming while the water is freezing. This interruption of the molecules' ability to bind at freezing temperatures requires lower temperatures to achieve a solid state, a process called *freezing point depression*. While the extracellular fluid can freeze at around 30°F, the intracellular water typically freezes at around 14°F. As the temperature of meat drops below freezing, the extracellular fluid begins to form solid ice crystals. Under normal home or commercial freezing environments (temperatures above −40°F), these ice crystals develop quite slowly, growing larger as they form a network.

This is the beginning of freezing any tissue: all the extracellular fluid needs to freeze before the intracellular fluid can freeze. But, as the extracellular ice crystals are forming, they start to leach moisture out of the cells, further concentrating the substances within the intracellular moisture — those same substances that interfere with water's ability to freeze — thus causing the freezing point inside the cells to drop even further and requiring

---

THE RIGHT FREEZER

Before you harvest meat, make sure that your freezer space is adequate. In most cases, a stand-alone freezer will be the best solution. Stand-alone models provide better temperature accuracy, more space, and slightly shorter freeze times than the freezers provided in conjunction with a refrigerator. This all adds up to better meat preservation. The right size can be hard to estimate — you need enough space for all the meat and then some to help organize and access everything. A helpful estimation is 50 pounds of meat per 2 to 3 cubic feet of freezer space. Chest freezers offer an efficient use of space, though a well-organized stand-up freezer will make the packaged meat more accessible.

an even lower temperature to achieve a truly solid, preserved state. Furthermore, the extracellular absorption of water dehydrates the cells, the first of many ways in which freezing damages meat.

The Importance of Rapid Freezing

Water in meat is held within fragile structures like cellular membranes. As the water freezes and expands, the sharp ice crystals cause the cellular walls to rupture. The slower the water freezes, the more expansion occurs and the sharper the ice crystals are, resulting in greater damage to these water-holding frameworks. To make matters worse, the crystals continue to grow after the product appears to be completely frozen, drawing more moisture out of the cells and causing more dehydration and damage.

The ice acts as a plug to the holes made in cellular walls and other structures, but as meat thaws some of the water seeps out of these holes, draining nutrients, pigments, and minerals along with it, resulting in what is called *drip loss*. You may have experienced this: thaw a frozen steak on a plate, and there will be a pool of red liquid collecting around it that you may have thought was blood. It is not: rather, it is formerly frozen liquids full of nutrients and proteins like myoglobin, which provides the red hue. Drip loss is inevitable, and all frozen products will experience it; the amount of drip, and the resultant quality of the meat, is directly linked to the speed of freezing. Drip loss is not the only quality risk from the formation of large ice crystals. These structures will also break emulsions — the stable mixtures of pureed substances, like mayonnaise, hollandaise sauce, or hot dog filling — causing the separation of fats and proteins. This is the reason why hot dogs and other emulsified meat products are not recommended for freezing. Commercial hot dogs destined for the freezer utilize additive ingredients to counteract this effect and stabilize the emulsion.

Because of its effect on the quality of the meat, the speed of freezing is one of the most important factors in preservation. Commercially frozen foods are exposed to extremely low temperatures, −30°F and lower, and high air velocities that promote rapid freezing and thus smaller ice crystals. At home the best option is a stand-alone freezer that provides a quick-freeze or deep-freeze option, or another comparable setting that can be used temporarily when initially freezing items.

KILLING PATHOGENS

Slow freezing is detrimental to more than just the cellular structures of the meat. Many bacteria and other microbes, especially those that are single-celled, can be destroyed during freezing; however, because the numbers would be uncertain and the effect not uniform, it is safer to treat freezing as a method of suspending microbial functions rather than destroying them.

AVOIDING DEHYDRATION

The formation of large ice crystals may be one of the more damaging effects of freezing, but it is not the only risk to consider. Freezer burn is simply a form of dehydration and oxidation caused by exposure of a frozen product to air. It is a consequence of improper temperature maintenance or packaging and can severely reduce the quality and nutrient value of frozen products. While freezer burn most often does not make the product unsafe to consume, it does make it unpalatable.

> Large ice crystals that form during freezing will break emulsions — the stable mixtures of pureed substances, like hot dog filling — causing the separation of fats and proteins. This is why hot dogs and other emulsified meat products are not recommended for freezing.

Dehydration occurs in a frozen product despite the fact that water is held in a solid state. The air within a freezer is innately dry; when it comes into contact with frozen moisture, a process called *sublimation* occurs, in which the solid ice crystals change state to gaseous water vapor without transitioning through the liquid state. Sublimation pulls moisture from the interior of the frozen product to the surface, where it forms large ice crystals and causes the area to permanently dry out, essentially freeze-drying it. The ridges and holes left by the absence of water reflect light, causing light-colored spots on the meat. Furthermore, the air brings oxygen into contact with the product surfaces, causing oxidation and the resultant discoloration to darker or brown tones. (For more on myoglobin and oxidation of meat and fats, see page 14.)

The Tenets of Freezing

Following some baseline recommendations can help you avoid the more detrimental consequences of freezing.

MAINTAIN A CONSTANT TEMPERATURE

First and foremost, keep frozen foods at 0°F or lower and avoid fluctuations in temperature. Fluctuations can result from opening the freezer too often or from inaccurate temperature regulation by the freezer itself. Changes in temperature can cause additional growth spurts in ice crystals, which in turn will contribute to additional cellular membrane damage, drip loss, and moisture loss. If you suspect that your freezer doesn't maintain a consistent temperature, set it lower than 0°F to prevent harmful fluctuations.

DON'T USE FREEZING AS A FORM OF SANITATION

Freezing does nothing to improve the quality of meat; it merely aims to maintain the quality of the product as it was prior to freezing. In fact, meat that has been frozen will be of lesser quality than fresh meat (though with the right approach and equipment, the degradation of quality will be slight). Therefore, start with the highest-quality product possible. Do not try to save meat that is beginning to spoil by freezing it. Freezing is not a form of sanitation; you cannot expect it to reduce the colonies of microbes that can cause spoilage or foodborne illnesses.

PORTION BEFORE FREEZING

Before you freeze them, portion your meats according to your expected needs — single pounds of ground meat, individual steaks, small roasts, and so on. When freezing multiple portions together in one package — say, a few steaks or some preformed burger patties — prevent unappetizing discoloration by making sure the cuts don't touch one another. Avoid having to thaw and refreeze portions, because the adverse effects will become more pronounced with each freezing.

Do not freeze whole primals or large cuts that will not be cooked whole. The larger the item, the slower the freeze, and the higher the risk of cellular damage.

FREEZING ORGANS

Organs — such as the liver, the stomach, and the intestines — are the operational center of an animal's existence. While all muscle tissue bears a likeness, organs are drastically dissimilar from one another due to their highly specialized functions. This makes for a splendid culinary challenge, offering a wide range of textural and flavor possibilities far beyond that of muscle meat.

Freshness with organs is paramount. Organ tissues tend to have higher levels of enzymes to aid in their activity, and for that reason they degrade quickly. Muscle meat may benefit from aging, but organs immediately begin to deteriorate once the systems shut down that have controlled enzymes and maintained balance in the body.

Freezing organs is the best way to halt the degradation of quality, and doing so immediately after slaughter is ideal. This includes all visceral organs, tongues, and marrow bones (which are organs, too). Use an ice-water bath to rapidly bring down the temperature of organs. As soon as they're cold, proceed with packaging and freezing. You can also use an ice glaze to package them. (See page 322 for more information on ice glazing.)

Freezing time is squared in relation to thickness — cut a rib steak twice as thick as another, and it will take four times as long to freeze. Smaller packages of meat will freeze quicker and therefore maintain quality for longer periods of storage. When simultaneously freezing multiple items for the first time, disperse them among the shelves of the freezer to promote quick, even freezing. Use the quick-freeze or deep-freeze option if the freezer has it. Do not stack or crowd items that are being initially frozen; all items need cold airflow to hit every surface to maximize results.

LEANER CUTS STORE LONGER

Fat oxidizes and goes rancid more quickly than lean meat does. Thus, leaner cuts of meat hold their quality longer than fattier cuts. If you expect extended storage, trim exterior fat to the thinnest acceptable amount prior to freezing. Unsaturated fats go through oxidative and deteriorative changes more quickly than saturated ones. Therefore, pork and poultry, which both have relatively high levels of unsaturated fats, will remain in palatable states for shorter periods than beef, lamb, and goat will. Ground meats will spoil more quickly than whole pieces of meat because they have more surface area and are therefore more susceptible to oxidation and other chemical changes.

COOKED AND CURED MEATS DON'T FREEZE WELL

Cooked and cured meats freeze poorly compared to fresh, uncooked meat. This is because cooking causes considerable moisture loss; with less moisture in the meat, the efficacy of freezing is also lessened. Furthermore, cured and processed meats have increased levels of salt, a mineral that acts as a catalyst for oxidative reactions, thus increasing the rate of rancidity.

Packaging

PACKAGING IS A LINE OF DEFENSE for your meat, shielding it from the detriments of freezing. If packaged improperly, your meat will lose its battle against ice and oxidation, and will be dry, limp, and unpalatable. With the proper tools and approach, however, preserving meat with little to no perceivable degradation is entirely possible.

The Tenets of Packaging

Packaging for freezing has one objective: to preserve the quality of the product for as long as possible. Each decision about which materials to use and how to use them should be made in response to one question: Will this help preserve my product as long as I need it to? Technology in packaging has developed immensely over the many decades since commercial and home freezing became the norm, but some simple rules based on scientific principles have gone unchanged. And, thankfully, these protocols apply to storage for both fresh and frozen goods.

AIRTIGHTNESS

Packaging intended for the freezer needs to be airtight — both moisture-proof and vapor-proof — in order to have any chance of preserving meat. Permeable materials allow air and moisture through, resulting in freezer burn and quickened rancidity. This is why you never want to freeze meats bought from stores using the traditional "store wrap" — a foam tray with a drip-liner wrapped in plastic — a style of wrapping that intends for air passage in and out, allowing the meat to "bloom," the process in which the myoglobin, the protein responsible for meat's coloration, oxidizes and turns bright red. (For more on oxidation, see page 14.)

EXPULSION OF INTERIOR AIR

While being impervious to the outside air and moisture, packaging should also promote expulsion of as much interior air as possible. Skintight packaging is ideal, with the exterior material touching the surface of the meat wherever possible. This prevents air from reaching the surface. Pliable materials are the best option for nonliquid products, and those with good tensile strength will enable tightening around the meat when air is expelled. Any applicable material should also be easy to remove from the meat after thawing.

DURABLE, FOOD-GRADE MATERIALS

Materials need to be made from an odorless food-grade substance that is tasteless and resistant to transferring any off-flavors. Its impenetrable state should also prevent the transmission of off-flavors between products frozen within the same space. (The last thing you want is for your steak to taste like its plastic container or the whole frozen fish next to it.) Cracking, brittleness, or other forms of detrimental warping should never occur during freezing, when moisture is expanding, or when the product is kept at low temperatures indefinitely. You will likely need to reorganize your frozen products from time to time; therefore, the packaging will need to protect the interior product while also being durable enough to resist the resulting wear and tear. When applicable, package designs that enable convenient storage, stacking, and organization are preferred.

PROPER LABELING

Create labels that you can read easily without having to hold the item in your hand. Otherwise you will be riffling through the entire freezer, picking up item after item, searching for a specific cut. Most meat packaging will be disposable, so the label may simply consist of information written directly on the package with a high-quality permanent marker. But tagging on white or brightly colored backgrounds certainly improves legibility. Temporary labels need to have a freezer-proof adhesive. Reusable containers, used mostly for liquids, should have the space for proper temporary labeling.

AFFORDABILITY

Packaging needs to be affordable, falling into range with your budget. Expensive solutions abound, but you can also find affordable ones. While protecting your product is important, so is profitability. And those with customers must use packaging that appeals to buyers' needs.

Packaging Solutions

Currently there is no single type of packaging that is both perfect and affordable, though some types come very close. The solutions here range from ideal to bare-bones. Assess your priorities by considering both your products and your preferences. Depending on the product to be packaged, the best solution may be to employ multiple techniques.

VACUUM SEALING

Vacuum sealing is by far the single best option for packing perishable meat products. It's the only method that's applicable to every animal species and works in all conditions. Interior air is removed under pressure, pulling the pliable bags tight against the product surface. The bags, made from impermeable food-grade plastic, are sealed permanently through the application of heat and can be used for products in any shape or size. Plus, the process is quick and easy to execute.

Vacuum sealing does have its drawbacks, however. First, investing in a vacuum-sealing machine that can adequately handle your volume may tip the cost scales. Second, edge sealers — the more affordable models designed for home use — do not handle liquids or highly moist products well (this does not include raw meat). Third, home models are also unable to handle extra-large items like a standing rib roast of beef. Fourth, vacuum-sealing bags tend to be more costly than other options, even in bulk, and are prone to being punctured, allowing dry air to flood the interior and leading to freezer burn (though it may be minor). Despite these caveats, vacuum sealing is still the best solution for many processors, especially those dealing with multiple species.

EDGE SEALERS. There are two types of vacuum sealers: edge sealers and chamber sealers. The most common models for home use are edge sealers, with the most popular being the Food Saver brand. The operation requires much simpler machinery than the chamber sealers, making them more affordable and smaller in overall size. Their

Before storage, ensure you that your vacuum-sealed bag is well sealed and no protruding bones present a risk of puncturing the bag, which will inevitably lead to freezer burn and spoilage.

major drawback is the inability to handle liquids or highly moist products.

Every model will have specific instructions for use, but the concept is the same across the board. Start with a bag, either premade or cut from a roll, that is large enough to fit the product while allowing enough space for sealing. Place the product inside, insert the bag into the machine, and start the vacuum. It will draw out the air, compress the pliable bag around the product, and seal the bag.

CHAMBER SEALERS. The industrial-oriented machines for sealing foods are known as chamber sealers. Rather than suck the air out of the bags, as with edge sealers, chamber sealers utilize a different technique. The bag, with food inside, is placed into the chamber with the opening of the bag laid across a sealing strip. When the chamber is closed, air is pumped out of the chamber, thus creating a low-pressure environment, or vacuum, inside. Once the target pressure is reached, the bag is sealed and the pressure from within the chamber is released. Since the bag was sealed when there was almost no air inside the chamber, once the chamber is opened, the bag compresses tightly around the food. This allows for the bag to contain liquids. Chamber sealers also allow you to control the amount of compression, which is helpful when working with products that need to hold their shape (cased sausages). Compression also helps to speed up marinating: when meat and liquid are sealed together, the pressure forces the marinade into the meat.

FREEZER-GRADE WRAPPINGS

Manually wrapping items is another effective approach, and there are many kinds of materials available for use by the home processor. The main challenge in manual wrapping is achieving a skintight, airtight package. The most suitable choice is a moisture-impermeable plastic wrap. The elasticity of plastic wrap enables it to conform to various shapes, allowing for a closer fit, thereby expelling more air. This material is often cheaper than any bag option, especially when you buy it in bulk. But there is a trade-off: the need for another layer. You'll need to use freezer or butcher paper in conjunction with plastic wrap to provide the necessary protection against puncturing. Any suitable paper will be two-layered, the exterior being paper and the interior being a thin plastic film to retard air.

This approach works well for small items such as portioned cuts, small roasts, or ground meat, but it is cumbersome and less effective when dealing with whole subprimals. Also, wrapping with plastic and paper is a two-step approach that takes more time than a single-step solution. For products without a uniform shape, appearance tends to suffer, because the ability to wrap well with paper is challenging. In turn, salability is also affected, potentially from appearance but also because the interior product is obscured by the opaque outer wrapping. In the end, the main advantage to manual wrapping is cost savings in materials (despite the need for two layers) and the lack of equipment needed.

TIPS FOR BETTER VACUUM SEALING

- To cut down on costs, consider joining with other processors and farmers in purchasing a vacuum-sealing machine for shared use.

- To avoid punctures, use additional padding for bone-in butchered parts and other items that have bony or sharp protrusions. You can also use cut-up pieces of vacuum bags or prefabricated bone guards to cover the areas of concern, effectively creating a double layer of plastic, before you seal the bag. For very sharp areas, consider taking a sanitized rasp or file and smoothing the edges.

- While any size bag can be used for any size item, reduce costs by having a selection of bag dimensions, using the smallest option for the product being sealed.

- Before filling a bag, fold the edge down about 3 inches (typically the area that gets melted into a seal) in order to avoid soiling it with fats or getting it wet, which will prevent a clean seal.

TECHNIQUES 101

How to Store Meat with an Edge Sealer

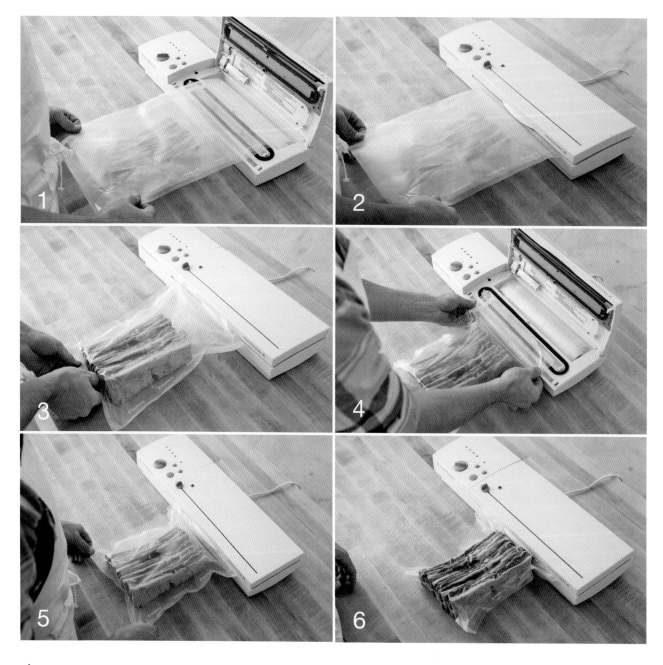

1. Measure out the length of bag needed to accommodate the contents.

2. Cut the bag from the roll and seal one end.

3. Fold down the edge of the bag and place meat inside.

4. Unfold the bag edge and align across the sealing strip with the open end of the bag inside the suction chamber.

5. Close the lid and start the removal of air from the bag.

6. When the indicated level of vacuum is achieved, the unit will permanently seal the bag.

TECHNIQUES 101

How to Store Meat with a Chamber Sealer

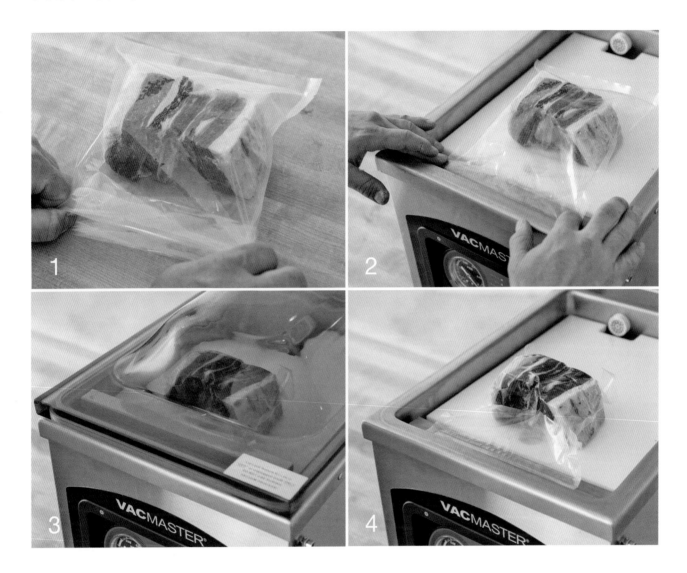

1. Fold down the edge of the bag and place meat inside.

2. Unfold the edge of the bag and place in the chamber. The edge of the bag should lie flat against the sealing strip with no wrinkling.

3. Close the lid, which starts the removal of air from within the chamber.

4. The lid opens after the bag is sealed and the chamber returns to normal pressure.

TECHNIQUES 101

Two Methods for Wrapping Meat

THE DRUGSTORE METHOD

1. If freezing meat, consider first wrapping it in plastic wrap. Place meat near the corner of the paper nearest you. If freezing meat, consider first wrapping it in plastic wrap.

2. Lift the top and bottom edges and align them above the meat.

3. Starting at the aligned edge, fold the paper down, in 1-inch increments, and crease tightly. Repeat until the paper is tight against the meat.

4. Fold the four sides of the paper inward,pressing out the air and forming a triangular point on both ends.

5. Flip the package over and bring the triangular points together in the middle.

6. Secure the points with freezer-grade tape and label the package.

TECHNIQUES 101

THE BUTCHER METHOD

1. If freezing meat, consider first wrapping it in plastic wrap. Place meat near the corner of the paper nearest you. If freezing meat, consider first wrapping it in plastic wrap.

3. Flatten the sides of the paper and crease tightly, expelling the inside air.

5. Keeping the paper taut, roll the meat toward the far corner of the paper.

2. Fold the near corner of the paper over the meat and tuck underneath your cut.

4. Fold the sides of the paper up and tightly across the meat, further expelling air.

6. Secure the final corner of the paper with freezer-grade tape and label the package.

> Ice conducts heat up to four times faster than liquid does. Because of this, when the exterior of frozen meat thaws first, it actually insulates the frozen center, preventing the warmer exterior temperatures from reaching the interior.

ICE GLAZING

Another effective way of keeping air and other oxidizing elements away from your meat is to lock them in ice, a process called ice glazing. Ice glazing is simple and incredibly effective. When the meat is encased in ice, the chance of oxidation is highly unlikely. Some sublimation may occur, but its effect on your edibles will be insignificant. This approach is best for organs. Anything larger tends to be too cumbersome when storing and thawing.

Find a waterproof container that's large enough to hold the items you wish to store while being covered with water. Organs, especially those of smaller animals, should be portioned and frozen in convenient amounts. Place the items in the container and cover them with water. Carefully place the container in the freezer and loosely cover it, allowing some air to escape (to avoid the top popping off). Return after 24 hours and tightly seal the container for long-term storage. Ice-glazed items should be thawed like any other frozen cuts, though it is advisable to have something beneath the thawing container, as cracks may have developed in the container during the freezing and storing process.

ZIP-TOP BAGS

Last, and least desirable, is the zip-top plastic bag method. Using a zip-top bag will inevitably cause some freezer burn, the severity of which is dependent on how long you store the frozen meat. Therefore, use zip-top bags only for short storage times — days to weeks at most — and take the effort to remove as much air as you can. Many companies offer bags with a valve, allowing for the removal of air with a hand-pump or other device. These are ideal for those not wanting to invest in a bona fide vacuum sealer. And, many of the bags are resealable and reusable, to boot. You can also use a simple bucket or stock pot full of water. Take the bagged product and slowly lower it into the water, being careful not to let the opening of the bag fall underneath the surface of the water. The water will force out a majority of air as the bag descends. Stop at the lowest point at which you can still seal the bag above the waterline. Once the bag is sealed, remove it from the water, dry it off, and store it.

Additional Materials

The preceding solutions can all benefit from some additional materials, depending on what is being packaged. Freezer sheets, or another type of liner, are required when packaging portioned or separate red-meat items together in a package. Foam trays can help provide structure to items, especially loose products like patties or cubed meat and those being stored with manual wrapping. Soaker or purge pads, often accompanying foam trays, will absorb the drip coming off thawing items, helping to prevent any

---

FOILING THE ELEMENTS

Aluminum foil can be an alternate option for the initial layer of wrapping, but it is hard to get a hermetic seal with foil. Plus, it tears easily. But aluminum foil is an excellent choice as an extra measure against oxidizing elements. After wrapping with plastic wrap, cover the cut with a single layer of aluminum foil, carefully sealing the edges. Then proceed with wrapping in paper. This layer of extra protection is often helpful for cuts with sharp protrusions (like bone-in cuts), whole subprimals, primals, or other large items.

off-flavors that may develop. Ground-meat bags are great for that item and come in sizes demarcated for weight, making it easy to estimate the volume when stuffing them. For all bags, the opening must be sealed; the three options for that are staples, hog rings, and bag tape. The first two are the best all-around options, while the bag tape is really only effective with ground-meat bags.

Get a scale if recording weight is a priority. Prices range dramatically, but something as simple as a bathroom scale will do or, for more precise results, a digital model that has a decimal readout. The scale will need to accommodate the maximum size that you will be processing.

Labels for home use should be large enough to contain such information as contents, weight, and date. For salable items labels may need to include further information such as price, breed, or information about your farm. Most packages of meat cuts are disposable, allowing you to write simple information directly on the package with a high-quality permanent marker (although writing on white or brightly colored backgrounds certainly improves legibility). Temporary labels on reusable containers need to have a freezer-proof adhesive.

Thawing

DO NOT UNDERESTIMATE the importance of proper thawing. You will spend a lot of time on butchering, packaging, and freezing the bounty from a harvest, and your efforts will be wasted if you thaw the meat incorrectly. Poorly thawed meat is, at best, poorly textured and, at worst, dangerous to consume.

What *Not* to Do

Never defrost a product by microwaving it, sitting it in a warm-water bath, or leaving it to sit out on the counter or anywhere else that has warm ambient temperatures. Microwaving may be a safe way to quickly defrost meat, but the result is a steamed, partially cooked mass that has already undergone major chemical and physical change, none of which is advantageous to texture or taste. Attempting to speed up the defrosting process by exposing the meat to any temperatures above that of your fridge is just dangerous. These temperatures inevitably reside between 40°F and 140°F, the range considered the danger zone. (For more on the danger zone and foodborne illness, see page 23.) This is especially applicable to poultry and offal, though the danger is inherent in any perishable product.

One thing to understand is that thawing meat takes longer than freezing it does; there aren't any suitable shortcuts. The reason for this has to do with the comparative ability of water to transfer heat in different states. Ice conducts heat up to four times as fast as liquid water does. During freezing, the exterior of meat solidifies first, which helps conduct heat out of the meat, quickening the rate of the temperature drop inside. Thawing acts in just the opposite way: the exterior thaws first and, due to less efficient conductivity, actually insulates the frozen center, preventing the warmer exterior temperatures from reaching inside. It is for this reason that the time needed to thaw meat properly is considerably longer than that needed to freeze it.

Proper Thawing

The ideal methods of thawing just take some forethought and time. First, and easiest, is the refrigerator. Anything can be safely thawed in a properly working fridge, but it takes a while because air doesn't conduct heat well. Thus, the air that surrounds the frozen package is an inefficient coolant. Nonetheless, it works. Leave the meat packaged, set it in a container to catch any drip loss or condensation, and allow at least a day for small items, like steaks and sausage, and up to three or more days for large items like roasts and whole subprimals.

The second method, using ice water, illustrates how much more efficient a liquid is than a gas at conducting heat. This is the only suitable shortcut for thawing. Use a container about four times the size of the frozen product. Leave the product in its package and submerge it in cold water. (I also add some plates on top to keep the item submerged, but anything with weight will work if you find the product bobbing up for air.) Add lots of ice and leave the water trickling. The ice-cold temperature will actually improve the thawing time while also retarding microbial growth. The trickling flow of liquid will keep a mild circulation in the water bath, helping keep cold water against the frozen product at all times. Using this method, you can defrost small items in as little as 30 minutes and large items in a couple of hours (or more depending on how large).

RIB AND LOIN STEAKS

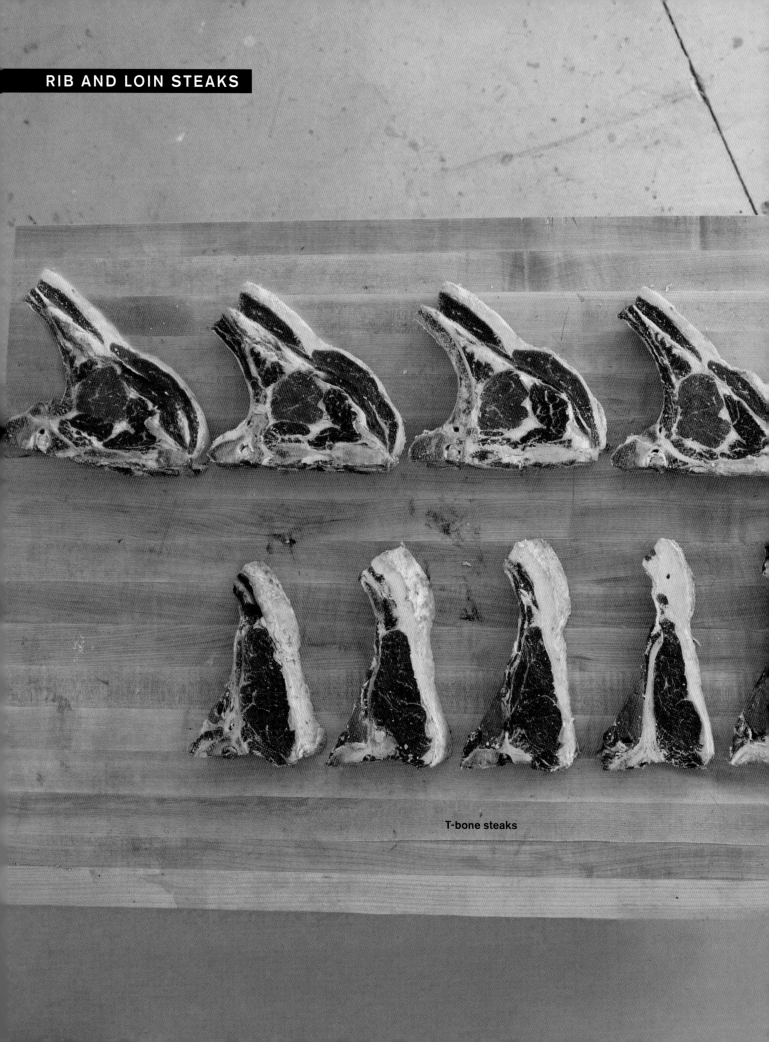

T-bone steaks

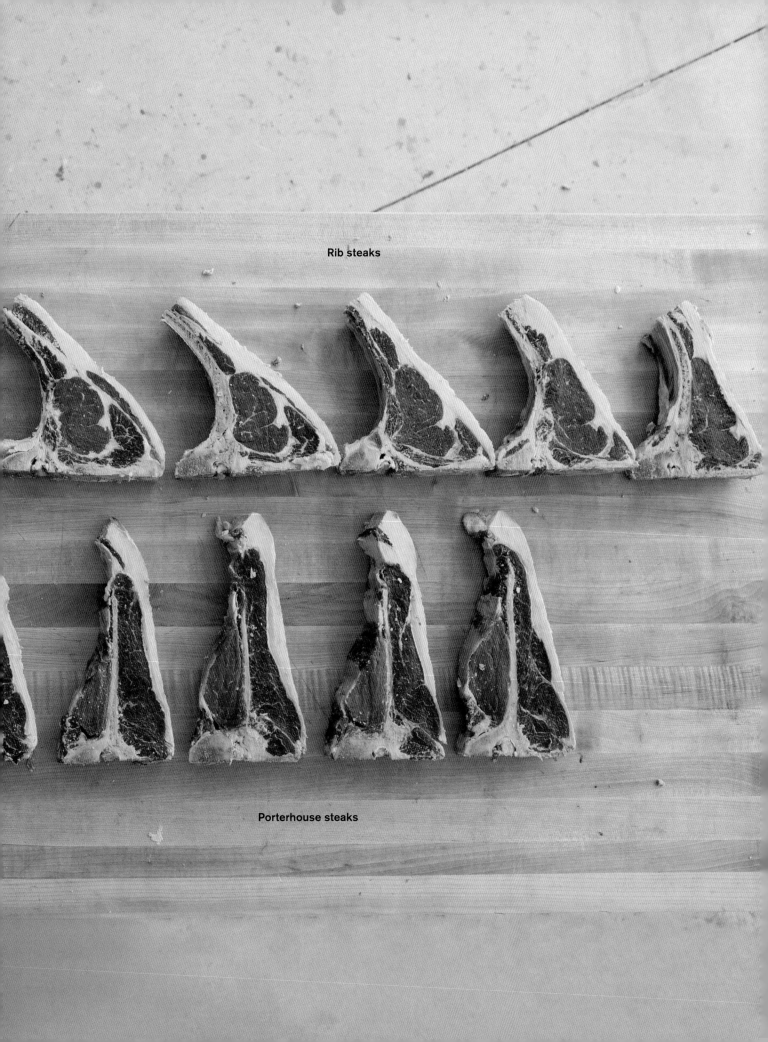

ACKNOWLEDGMENTS

T**HERE ARE SO MANY PEOPLE** that I feel blessed to have worked with, during my career as a butcher and the creation of this book. I don't have space to thank them all, so here's the "short" list, starting with my dear editor, Carleen. She had no idea what she was stepping into when she contacted me the first time, God bless her. Even so, she's owned this project from the onset and stood behind it, and me, thus allowing it to maintain more of my original vision than I ever expected. It simply wouldn't be the project it is without her guidance and support. It's been a true joy working with her. (I think it's toughened her up a bit, too.)

Carolyn Eckert may be the warmest, most dedicated, and talented designer I've worked with. I had immediate confidence in her ability to produce designs that aesthetically and functionally worked (even for me!). What a relief. Her positivity is infectious. And the folks at Storey Publishing, who surprised me all along the way with their openness and sincere dedication to publishing books that educate. Their decision to turn this project into two volumes (this book and its companion, *Butchering Beef*) was visionary. They listen to their authors, and really hear them. And, of course, no book published in this modern era could be what it is without the talented and dedicated work of a photo retoucher. Hartley Batchelder, your work on this project was indispensable.

Joe Keller and his assistant, Jeff, were the perfect choice for photographing this project. It takes a certain kind of person to stand in a cold rain all day capturing images of slaughtering without losing his sense of humor. Not only are his images wonderful, but he is as well.

William P. Barlow (a.k.a. Billy the Butcher) was generous enough to volunteer his help on consecutive days of our photo shoot. A chef, butcher, consummate ice cream maker, and dear friend, Billy's understanding of food and his brilliance with it as a medium is stunning, and I'm indebted to him for innumerable areas of education within this industry.

I don't think we would have ever survived our first day of slaughter photography without the help of Greg Stratton. Proficient, efficient, and easy to work with, Greg's knowledge and foresight helped us avoid numerous slip-ups as the rest of the crew was focused on "getting the shot." I am truly in his debt.

Eric Shelley surprised me from the moment I met him; his depth of knowledge is extraordinary and his generosity in sharing it even more so. His compassion for animals is evident in the way he talks and teaches, and is demonstrated when you see him handling livestock.

I owe my foundational knowledge to his talent as an instructor and mentor. Clint Lane — the Robin to Eric's Batman — is also a fantastic teacher. He makes it look too easy.

Speaking of those who make it look too damn easy, let's talk about Jose Morquecha, butcher and pasta chef at Blue Hill at Stone Barns. His innate abilities are staggering, and I've benefited immensely from the opportunities I've had to work at his side. Watching him cut is like watching the slow, fluid strokes of a painter, though before you know it, it's over and you're scratching your head, wondering how he finished so quickly. His generosity, in general but also with his butchering wisdom, is why I'm indebted.

There are so many talented butchers cutting meat today — too many to mention — and they all inspire and push me to better understand the utility of these animals. I can learn something from each one of them, and look forward to gleaning such knowledge when our paths cross.

I truly appreciate the opportunity created by Severine von Tscharner Fleming, leader of a fabulous and influential collective of young farmers called the Greenhorns. Look 'em up.

We are all indebted to this country's farmers, small and large, who provide us with the necessary nourishment we crave. Connect with some in your area and show them the respect they deserve, and find out first hand where some of your local food is grown and raised.

A special thanks to the manufacturers of excellent apparel and equipment that enable us butchers to perform at the levels that we do, especially the companies that were featured in this project: Rabbit Wringer, R. Murphy Knives, and Duluth Trading Co.

My mother instilled this love for and fascination with food at an early age; I was "making" dinner for my family before I could read and write. That allure with what we consider sustenance has never stopped. The greatest gift of that is simple: I know how to cook. I've lived a happier, healthier life because of it, thanks to her. Her love and my father's support have given me the confidence to continue expanding my comfort zone through risks and adventure.

My daughter, Moxie Blaze, challenges me to better understand myself while providing joy and love every day. She was born right when this project began. Whoa. Amazingly enough, I was able to write these books from home while also helping to raise a newborn, thanks to my loving wife, Lola. She understood the challenge of balancing these priorities and helped run the house when I was under the gun. Hence, this project, as with any book, is a burden largely shared by the partnership (in so many ways). I couldn't have done this without her.

BIBLIOGRAPHY

BOOKS

Chambers, Philip G., Temple Grandin, Gunter Heinz, and Thinnarat Srisuvan. *Guidelines for Humane Handling, Transport and Slaughter of Livestock*. FAO, RAP Publication, 2001.

FAO. *Guidelines for Slaughtering, Meat Cutting, and Further Processing*. FAO, 1991.

Heinze, Gunter, and Peter Hautzinger. *Meat Processing Technology for Small- to Medium-Scale Producers*. FAO, RAP Publication, 2007.

Lawrie, R. A. *Lawrie's Meat Science*. CRC Press, 1998.

McCracken, Thomas O., Robert A. Kainer, and Thomas L. Spurgeon. *Spurgeon's Color Atlas of Large Animal Anatomy: The Essentials*. Wiley-Blackwell, 1999.

McGee, Harold. *On Food and Cooking: The Science and Lore of the Kitchen*. Scribner, 2004.

Myhrvold, Nathan, Chris Young, and Maxime Bilet. *Modernist Cuisine: The Art and Science of Cooking*. The Cooking Lab, 2011.

North American Meat Processors. *The Meat Buyer's Guide*. North American Meat Processors, 2011.

Romans, John R., William J. Costello, Wendell C. Carlson, and Marion L. Greaser. *The Meat We Eat*. Prentice Hall, 2000.

Underly, Kari. *The Art of Beef Cutting: A Meat Professional's Guide to Butchering and Merchandising*. Wiley, 2011.

ARTICLES

Baird, Bridget. "Dry Aging Enhances Palatability of Beef." Cattlemen's Beef Board, 2008.

Boles, Jane Ann, and Ronald Pegg. "Meat Color." Montana State University.

Bonhotal, Jean, Lee Telega, and Joan Petzen. "Natural Rendering: Composting Livestock Mortality and Butcher Waste." Cornell University, 2002.

Bowker, B. C., A. L. Grant, J. C. Forrest, and D. E. Gerrard. "Muscle Metabolism and PSE Pork." American Society of Animal Science, 2000.

Brooks, Chance, Ph.D. "Beef Packaging." Cattlemen's Beef Board, 2007.

Calkins, Chris R., Ph.D., and Gary Sullivan. "Ranking of Beef Muscles for Tenderness." Cattlemen's Beef Board, 2007.

CDC, FDA, USDA. "Foodborne Illnesses Table: Bacterial Agents." USDA, 2004.

FDA, Keith A. Lampel (ed). "Bad Bug Book: Foodborne Pathogenic Microorganisms and Natural Toxins." FDA, 2012.

Goodsell, Martha, and Dr. Tatiana Stanton. "A Resource Guide to Direct Marketing Livestock and Poultry." Cornell University, 2010.

Grandin, Temple, Ph.D. "Recommended Animal Handling Guidelines & Audit Guide: A Systematic Approach to Animal Welfare." American Meat Institute Foundation, 2010.

Grimes, Lauren M., and Chris R. Calkins. "Mapping Tenderness of the Serratus Ventralis." Animal Science Dept., University of Nebraska – Lincoln, 2008.

Hamilton, D. N., M. Ellis, K. D. Miller, F. K. McKeith, and D. F. Parrett. "The Effect of the Halothane and Rendement Napole Genes on Carcass and Meat Quality Characteristics of Pigs." Journal of Animal Science, 2000.

Kim, G. D., J. Y. Jeong, S. H. Moon, Y. H. Hwang, G. B. Park, and S.T. Joo. "Effects of Muscle Fiber Type on Meat Characteristics of Chicken and Duck Breast Muscle." Division of Applied Life Science, Gyeongsang National University.

King, D. A., M. E. Dikeman, T. L. Wheeler, C. L. Kastner, and M. Koohmaraie. "Effects of Cold Shortening and Cooking Rate on Beef Tenderness." Cattlemen's Day, 2002.

Maddock, Robert., Ph.D. "Mechanical Tenderization of Beef." Cattlemen's Beef Board, 2008.

Miller, Dr. Mark. "Dark, Firm and Dry Beef." Cattlemen's Beef Board, 2007.

———. "Optimizing Blood Collection." Meat Technology Update Vol. 4(3), August 2003.

Parish, Dr. Jane A., Dr. Justin D. Rhinehart, and Dr. James M. Martin. "Beef Grades and Carcass Information." Mississippi State University, 2009.

Purchas, Roger W., Dean L. Burnham, and Stephen T. Morris. "On-Farm and Pre-Slaughter Treatment Effects on Beef Tenderness." Massey University, New Zealand, 2001.

Savell, J. W., S. L. Mueller, and B. E. Baird. "The Chilling of Carcasses." International Congress of Meat Science and Technology, 2004.

Savell, Jeff W., Ph.D. "Dry Aging of Beef." Cattlemen's Beef Board, 2008.

Scallan, Elaine, Robert M. Hoekstra, Frederick J. Angulo, Robert V. Tauxe, Marc-Alain Widdowson, Sharon L. Roy, Jeffery L. Jones, and Patricia M. Griffin. "Foodborne Illness Acquired in the United States — Major Pathogens." Emerging Infectious Diseases Vol. 17(1), 2011.

Scanga, J. A., K. E. Belk, J. D. Tatum, T. Grandin, and G. C. Smith. "Factors Contributing to the Incidence of Dark Cutting Beef." Colorado State University, 1998.

Snyder, Jr., O. Peter, Ph.D. "The Microbiology of Cleaning and Sanitizing a Cutting Board." Hospitality Institute of Technology and Management, 1997.

Van Laanen, Peggy. "Safe Home Food Storage." Texas Agricultural Extension Service.

Wheeler, T. L., and M. Koohmaraie. "Prerigor and Postrigor Changes in Tenderness of Ovine Longissimus Muscle." Journal of Animal Science, 1994.

Wulf, Duane M. "Did the Locker Plant Steal Some of My Meat?" South Dakota State University.

WEBSITES

Gamble, H. Ray. "Trichinae: Pork Facts — Food Quality and Safety." USDA, Agricultural Research Service

http://www.aphis.usda.gov/vs/trichinae/docs/fact_sheet.htm

Grandin, Temple. "Recommended Stunning Practices." http://www.grandin.com/humane/rec.slaughter.html

USDA. "Using Dentition to Age Cattle." USDA. http://www.fsis.usda.gov/ofo/tsc/bse_information.htm

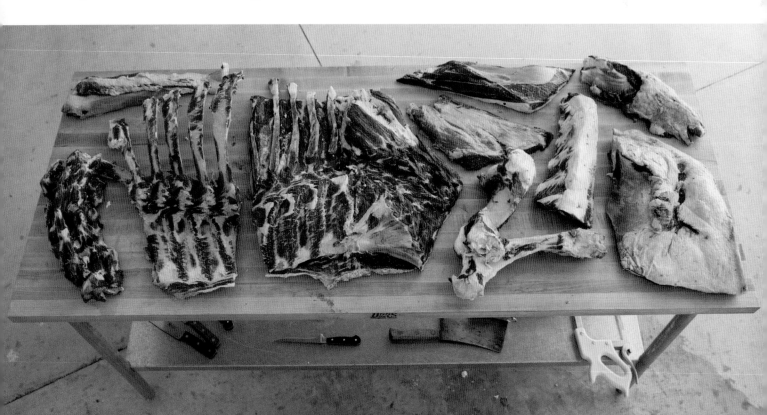

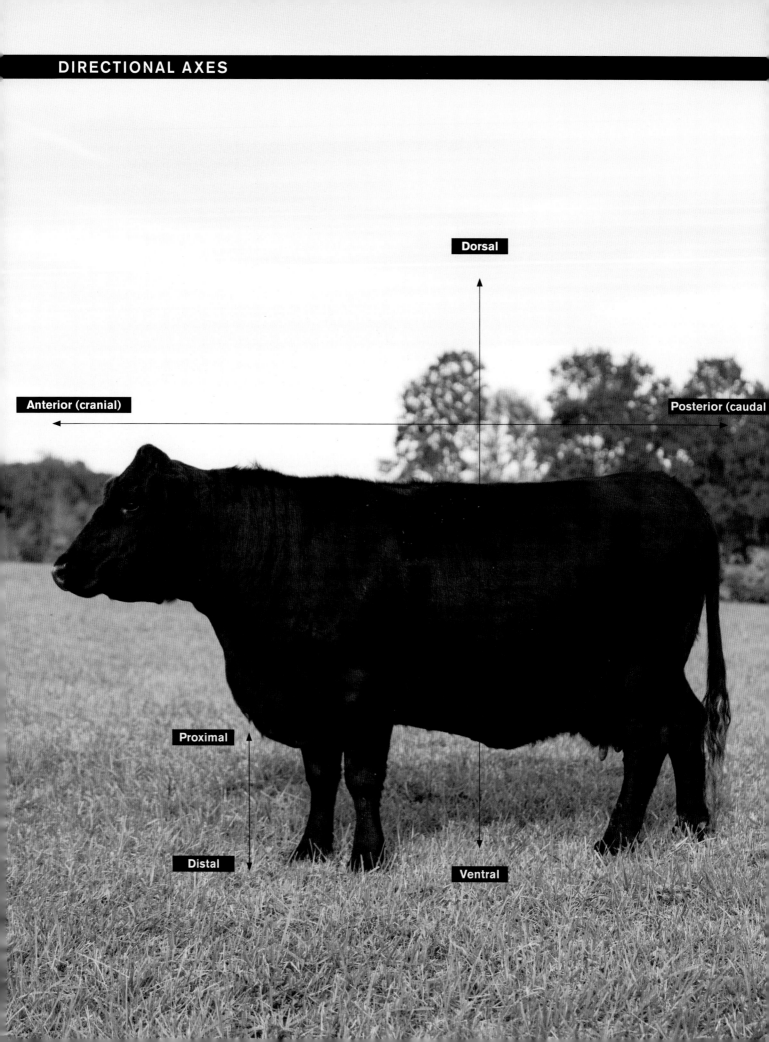

GLOSSARY

ABDOMEN | the area between the thoracic cavity and sacral region; the belly

ABDOMINAL | relating to the abdomen

ABOMASUM | the fourth, and final stomach, of the rumen before food is passed into the small intestines

ACETABULUM | the socket of the hip where the ball of the femur attaches

ACHILLES TENDON | the tendon connecting the shank muscles to the hock or cannon bones

ACTIN | one of the two main contractile proteins within muscle fibers; one of the proteins responsible for the effects of rigor mortis

ADIPOSE TISSUE | commonly known as fat

AGING | a process by which natural chemical reactions occur under controlled conditions in order to counteract the effects of rigor mortis and increase flavor and tenderness within meat

AITCHBONE | half of the pelvis after a carcass has been split into sides

ANTERIOR | relating to the front of an object; occurring closer to the head of the animal; the opposite of posterior

BALL OF FEMUR | the proximal end of the femur bone that attaches to the pelvis at the acetabulum

BILE SAC | see Gallbladder

BLADEBONE | see Scapula

BLOOMING | the act of myoglobin oxygenation that results in the cherry-red, or reddish, color of meat

BOVINE | of or relating to the species of cattle

BREAK | to segment a carcass into primal cuts

BUNG | the anus of the animal

CANNON BONE | the lowest portion of the leg of an ungulate, connecting the hock or shank and the hoof; bears most of the animal's weight when walking; is used for raising the animal after stunning during exsanguination

CAUDAL | relating to, or occurring near, the rear end of the animal

CAUL FAT | a lacework of fat and connective tissue extending over the stomach and intestines; often used during aging to inhibit excessive moisture loss

CERVICAL | relating to the neck of the animal; occurring near the neck

CERVICAL VERTEBRAE | the bones in the neck

CHINE BONES | the bones of the vertebrae after a carcass is split into sides

COD FAT | fat occurring in the groin area of steer

COLLAGEN | the main protein occurring in connective tissue, bones, cartilage, and skin

CONNECTIVE TISSUE | see Fascia

DISTAL | occurring away from the center of the carcass; the outer point; the opposite of proximal

DORSAL | occurring near the back of the animal; situated closer to the back

DRESSING PERCENTAGE | the percentage of the live weight that is left over after slaughter, exsanguination, and evisceration; the carcass weight divided by the live weight then multiplied by 100

ESOPHAGUS | a muscular tube that passes food from the mouth to the stomach; also called the weasand

EXSANGUINATION | the draining of blood from a living animal; death from bleeding

FASCIA | sheaths of fibrous tissue, mainly composed of collagen, that connect and wrap organs, muscles, and other parts of the body

FEATHER BONES | the spinous processes of the vertebrae after a carcass has been split into sides

FEMUR | the largest bone in the body, connecting proximally to the pelvis and distally to the tibia; informally known as the leg bone

FIBULA | the smaller of the two lower leg bones; bears little weight compared to the tibia, and in some cases is fused with the tibia

FINGER BONES | see Transverse Processes

FINISH | the final fattening of an animal before slaughter; the amount of fat on an animal when it's slaughtered

FOODBORNE | occurring in, on, or around food; contracted through the consumption of food

FOREQUARTER | the anterior half of a beef carcass

FORESADDLE | the anterior half of a veal carcass

FREEZER BURN | the drying out of meat through sublimation and oxidation of fats while in a frozen state

GALLBLADDER | an organ, attached to liver, that is responsible for bile production and excretion; requires removal before the liver can be considered edible

GAMBREL | see Achilles tendon; also known as gambrel cord

GELATIN | the resultant protein after hydrolysis of collagen

HINDQUARTER | the posterior half of a beef carcass

HINDSADDLE | the posterior half of a veal carcass

HOCK JOINT | the joint connecting the hock to the cannon bones or trotter; the joint at the distal end of the tibia

HUMERUS | the upper arm bone; connects ventrally with the radius and ulna and dorsally with the scapula

ILIUM | the anterior bone of the pelvis; one of the three bones that fuse together to form the pelvis or pelvic girdle; commonly known as the pin bone after a carcass is split into sides

INSENSIBILE | unconscious and unresponsive to pain; incapable of feeling, especially suffering

INTERCOSTAL | occurring between the thoracic rib bones

ISCHIUM | the posterior bone of the pelvis, also forming the dorsal edge of the pelvis; one of the three bones that fuse together to form the pelvis or pelvic girdle

LATERAL | of or near the side; the opposite of medial

LUMBAR | the area between the thoracic and sacral regions; the lower back

LUMBAR VERTEBRAE | the portion of the spine that comes after the thoracic vertebrae and before the sacral vertebrae; in lamb, goats, and pigs there are seven, while on bovines there are six

GLOSSARY CONTINUED

MEDIAL | near the middle or center; the opposite of lateral

METMYOGLOBIN | the oxidized state of myoglobin responsible for the brown hue of meat

MYOGLOBIN | the protein responsible for oxygen delivery from the bloodstream to the muscle fibers; occurs in various states depending on bonding with oxygen or other elements, resulting in the various states of muscle coloration; the default, deoxygenated state, is responsible for the purplish hue of meat that has not yet been exposed to air

MYOSIN | one of the two main contractile proteins within muscle fibers; one of the proteins responsible for the effects of rigor mortis

NECKBONES | see Cervical Vertebrae

OFFAL | the edible organs from the thoracic and abdominal regions of the body

OMASUM | the third stomach of the rumen and one of the sources of tripe

OSSIFICATION | the hardening and calcification of cartilage that happens as an animal ages; used to judge the age of some species

OXYMYOGLOBIN | the oxygenated state of myoglobin, responsible for the "bloomed" red color of meat

PATELLA | commonly known as the kneecap

PATHOGEN | a dangerous microorganism capable of causing disease

PAUNCH | an informal term for the stomach or rumen of an animal

PECTORAL | the region occurring across the lower portion of the thoracic ribs and sternum; the breast or chest region

PELVIC | relating to the pelvis or around the area of the pelvis

PELVIS | the fused combination of three bones: the ilium, ischium, and pubis; informally known as the hip bones

PH | the level of acidity within a solution

PIN BONE | see Ilium

PIZZLE | the penis of an animal

PLUCK | the thoracic viscera; the heart, lungs, and usually attached trachea

POLLED | an animal without horns, either naturally or by removal

POSTERIOR | the rear end of an object; occurring nearer to the tail or rear of something; the opposite of anterior

POSTMORTEM | after death

POTABLE | able to be safely consumed, as in clean drinking water

PRIMAL CUTS | the initial results from breaking down a carcass; the largest segments of a carcass that precede further breakdown into subprimals and portioned cuts

PROXIMAL | occurring near the center or attaching closer to the center; the opposite of distal

PSE | an acronym meaning pale, soft, and exudate meat; meat with too low a pH, occurring due to stress and excitement just prior to slaughter, making meat soft, mealy, and unable to properly hold moisture; occurs most frequently in pigs

PUBIC SYMPHYSIS | the cartilaginous midline between both sides of the pubis

PUBIS | the smallest of the three bones that fuse to form the pelvis, occurring nearest the groin

PURGE | moisture loss from fresh or cooked meats

QUADRUPED | an animal with four legs

RADIUS | one of the two lower arm bones, connecting proximally to the humerus; on most animals is fused with the ulna

RETICULUM | the second stomach of the rumen, occurring between the true rumen and omasum

RIGOR MORTIS | the chemical change after death in which actin and myosin enter into permanent and opposing states of contraction (ionic locking) causing a toughness in the muscles that lasts until the muscle structures deteriorate during aging

RUMEN | the first of four stomachs in ruminants; informally, the four-chambered stomach responsible for digestion in ruminants

RUMINANT | an animal with a rumen

SACRAL VERTEBRAE | the fused portion of the vertebrae occurring between the lumbar and caudal sections; also known as the sacrum

SCAPULA | the flat shoulder bone that connects distally to the humerus; also known as the shoulder blade

SHANK | the lower portion of an arm or leg, occuring above the hoof or foot

SIDE | to split a full carcass laterally in two; one half of a carcass that has been split in two laterally

SPINOUS PROCESSES | the bony extensions of the vertebrae

STERILE | containing no living organisms

STERNUM | the ventral-most bone of the chest, connecting both sides of the thoracic ribs; also known as the breastbone

STIFLE JOINT | the joint between the femur and tibia; also known as the knee joint

SUBPRIMAL CUTS | the breakdown of primal cuts into defined, subdivided sections; usually a collection of muscles separated through natural seams; can be further broken down into portioned cuts

SWEETBREAD | the thymus gland in adolescent animals; sometimes refers to the pancreas, which is also known as the "belly sweetbread"

SYNOVIAL FLUID | a lubricating fluid found amid joints to prevent friction; most frequently found in the stifle joint of beef

THAW RIGOR | the effects of rigor mortis occurring in a thawing carcass that was frozen before the onset and resolution of rigor mortis

THORACIC | the area between the cervical and lumbar regions; occurring in or around the chest region

THYMUS GLAND | a gland occurring in the neck that atrophies as the animal ages past adolescence; the main source of sweetbreads

TIBIA | the main lower leg bone, connecting proximally to the femur and distally to the hoof, foot, or cannon bones

TRANSVERSE PROCESS | the lateral bony extensions of the lumbar vertebrae; also known as finger bones

ULNA | one of two lower arm bones, connecting proximally to the humerus; often occurring fused to the radius

UNGULATE | a hooved animal

VARIETY MEATS | see Offal

VENTRAL | offering under or near the bottom of an animal or object; the opposite of dorsal

VERTEBRAE | the bones that make up the spine

VISCERA | the collective organs inside an animal

WEASAND | see Esophagus

RESOURCES

APPAREL
Duluth Trading Co.
Belleville, Wisconsin
866-300-9719
www.duluthtrading.com

BUTCHER BLOCKS
Green Mountain Woodworks
Talent, Oregon
866-535-5880
www.greenmtwood.com

John Boos
Effingham, Illinois
888-431-2667
www.johnboos.com

KNIVES
Friedr. Dick Corp.
Farmingdale, New York
800-554-3425
www.dick.de/en

R. Murphy Company, Inc.
Ayer, Massachusetts
888-772-3481
www.rmurphyknives.com

Victorinox Swiss Army Inc.
Monroe, Connecticut
800-442-2706
www.victorinox.com

Wenger NA
Orangeburg, New York
800-267-3577
www.wenger.ch/us/en

PROCESSING EQUIPMENT
Butcher and Packer Supply Company
Madison Heights, Michigan
248-583-1250
www.butcher-packer.com

LEM Products Direct
West Chester, Ohio
877-336-5895
www.lemproducts.com

MeatProcessingProducts.com
Incline Village, Nevada
877-231-8589
www.meatprocessingproducts.com

QC Supply
Schuyler, Nebraska
888-433-5275
www.qcsupply.com

Walton's, Inc.
Wichita, Kansas
800-835-2832
www.waltonsinc.com

SHARPENING
Edge Pro Inc.
Hood River, Oregon
541-387-2222
www.edgeproinc.com

FOODBORNE ILLNESS INFO
Bacteria and Viruses
U.S. Department of Health & Human Services
www.foodsafety.gov/poisoning/causes/bacteriaviruses

Estimates of Foodborne Illness in the United States
Centers for Disease Control and Prevention
www.cdc.gov/foodborneburden

Foodborne Illnesses & Disease Fact Sheets
Food Safety and Inspection Service, USDA
www.fsis.usda.gov/wps/portal/fsis/topics/food-safety-education

Partnership for Food Safety Education
www.fightbac.org

Safe Minimum Cooking Temperatures
U.S. Department of Health & Human Services
www.foodsafety.gov/keep/charts/mintemp.html

INDEX

A

abdominal viscera, removal of, 117–122
abscesses, 56
accessory muscles, 203
accidents, contamination and, 30
acidic tenderization, 65
adductor, removal of, 303
adrenaline rush, 17
age of animal
    meat tenderness and, 9
    muscles and, 7
    when to slaughter, 94–95
aging
    in a bag, 19
    dried tissue and, 56
    enzymes and, 18, 19
    hanging carcasses, 139
    length of time for, 18, 133
    in the open, 18–19
    veal, 128, 133
aitchbone, 97
    boning the sirloin and, 198
    removal of, 205
    round primal and, 137, 147
alkaline marinades, 65
anatomy/skeletal structure
    butchering cut points, 138
    reviewing, 60
    viscera and, 96–97
animal. *See also* pre-slaughter conditions; veal calves
    stress, 17
    well-being of, 2, 3, 5, 75
antibiotics, 21
aprons
    butcher coats and, 44
    protective, 45
atlas joint
    described, 97
    head removal and, 124, 125
attachments, severing, 59

B

back ribs, 238, 255
backstrap, 246
bacteria
    cold storage and, 32
    growth of, 25
    types of, 24–25
ball tip
    removal of, 201
    roast/steak, 259
bandsaw, 46
baseball/baseball cut, 278, 279
beef cattle. *See* animal

beef consumption, American, 93
beef cut sheet, 68–69
beef jerky, 290
bench scraper, 31, 42
bladder, 85
blade flap, 217
bleeding/blood drainage. *See* exsanguination
blood vessels, 97
"blooming," 14, 315
bone(s)
    chine bone, 245
    feather, 245, 255, 265
    finger, 196, 265
    making use of, 63
        marrow bones, 307
        stock bones, 306
    peeling, 62–63
    round, 63
bone buttons, 248
bone dust scraper, 42
bone-in rib roast, 244–47
bone-on rib roast, 248–49
bone saw/handsaw, 34, 38, 41, 51
boning
    basic, 60
    hot, 88
    joints, 60, 62
    peeling the bone, 62–63
    step-by-step, 61
boning hook, 35, 42, 136
    chuck tender and, 168
    marrow bones and, 307
    outward force and, 206
    scapula removal and, 172
    seam cutting and, 59
    sirloin flap and, 186
boning knife
    holding, 50
    as main knife, 36–38
bottom round
    long head/short head, 293
    round primal and, 147
    separating muscles, 208–9
bottom sirloin butt, seaming, 201
bovine spongiform encephalopathy (BSE), 28, 85
brains, 85, 97
breaking knife, 38, 39, 154
break joint, 138
brisket, 232–33
    boning, 233
    seaming, 234
    splitting, 105
brisket/foreshank primal, 141
    removal of brisket and foreshank, 159–160

    separation of brisket and foreshank, 161
    square-cut chuck and, 140
bruises, 56
BSE (bovine spongiform encephalopathy), 28, 85
bung
    sausage-making and, 85
    severing, 109–10
    tying off, 113
butcher blocks, 42, 43, 44, 136
butcher coats, 44, 45
butchering cut points, 138
butchering methods
    either/or decisions, 194
    two approaches, 150
    variations, 135
butcher knife, 38
butcher method, wrapping meat, 321
butcher's knot, 70–71
butcher's twine, 35, 39, 42–43. *See also* knots, making
butt, tenderloin, 198

C

calpains, 18
calves. *See* veal calves
cannon bones, 83, 96
cap complex, top round, 302
captive bolt pistols, 78, 90
captive bolt stunning
    insensibility and, 77
    penetrating/nonpenetrating, 78
    stunning spot for, 79
carbon dioxide, preservation with, 16
carcass
    breakdown, 139, 150
    cooling, 133
        hot boning and, 88
        quick rinse, 87
    hanging, 19, 77, 81–82
    hide removal and, 23
    quartering, 131–32
    rinsing, 130
    size, work surface and, 43
    skinning and, 101
    splitting evenly, 128–29
    transportation of, 56
    trimming, 130–32
    yield
        average, 93
        bones and, 63
        cut sheets and, 58
        maximum, 2, 57, 68

cardiac muscle, 11
carpaccio, 258
carpal joint, 96, 97
cartilage, costal, 138
cathepsins, 18
cattle. *See* animal
caudal vertebrae. *See* tail
caul fat, 119
chain
    removal of, 175, 272
    unlinking, 174
chamber sealers, 317, 318
cheek(s)
    masseter muscle, 86
    removal of, 126
chemical cross-linking
    connective tissue and, 7–8
    enzymes and, 18
    muscle tenderness and, 8, 9
chemical preservation, 16
chemical tenderization, 65
chine bone, 245
chops, veal, 51, 148, 212
chuck, 216–231
    under blade roast, 231
    blade flap, 217
    clod heart, 221–23
    cuts, common, 140
    eye, 218–19
        roll, 219, 231
        roll, seaming, 182–83
        roll, separating, 180
    neck, 224
    primal, 137, 140
    roll, 140
        boning, 173–183, 176–78
        seaming, 173–183, 179–181
    short ribs, 225, 238
    short ribs, chuck, 225
    shoulder tender, 226
    slow-twitch fibers and, 13
    splitting from rib, 154–55
    subprimals, 140
    tender, 200
        and shoulder clod, separation of, 165–69
    top blade, 227
cleanup. *See also* hygiene; sanitizing
    disposal of inedibles, 89
    post-slaughtering, 133
cleaver, 39, 40–41, 46, 47, 50, 136
clod heart, 221–23
clothing, 44–45
    aprons, 44
    basics, 91
    butcher coats, 44, 45
    gloves, 44, 45
    sanitary, 30
cold shortening, 17, 88
cold storage, 32

collagen
    bones and, 63, 306
    connective tissue and, 7
    cooking and, 8
    fibers and cross-links, 7, 8, 9, 12
    muscle tenderness and, 8, 9
    oxtail and, 309
    shank and, 299
complexus, separating, 182
composting, 89, 133
connective tissue
    fascia, 57, 60
    fats as form of, 9
    marbling and, 10
    muscle structure and, 6
    planes of, 2
    structure and, 7–8
    tenderness and, 12
    trimming, 36
consumption, beef, 93
containers
    edible, sausage-making and, 85
    for edible trim, 57
    for exsanguination, 80
    food-grade, 91, 316
    for inedibles, 56, 57, 117
    meat lugs, 47
contamination
    accidents and, 30
    avoiding, 83–84
    causes of, 23
    conditions for, 23–24
    cross-, 27
    evisceration and, 84
    foodborne illnesses and, 22, 29
    foodborne pathogens and, 22–23
    inedible trim and, 56
    microbes and, 22
    vegetables and, 23
cooked meats, freezing and, 315
cooking meat
    collagen and, 8
    muscle structure and, 6
    pathogens and, 33
    whole cuts, 33
cooling, 87, 88, 133. *See also* freezing; refrigeration
costal cartilage, 138
coulotte, 277
cow. *See* animal
cross-contamination, avoiding, 27, 32
cross-cut shanks, 300
cross-links. *See* chemical cross-linking
cured meats
    freezing and, 315
    myoglobin and, 15
cut points/skeletal structure, 138
cut-resistant gloves, 44
cuts, beef subprimals and, 214–15

cut sheet
    choices to make, 150
    form, example, 68–69
cutting boards
    sanitizing, 30–31
    wooden, 43–44
cutting tools. *See* boning knife; knives; tools

D

"danger zone"
    cold storage and, 32
    hot boning and, 88
    temperature range, 23–24
    thawing meat and, 323
dark, dry, and firm (DDF) meat, 17
"dark cutter" meat, 17
DDF meat, 17
death, testing for, 81
deep fascia, 60
defrosting meat. *See* thawing meat
dehydration
    frozen meat, 313
    products, 290
Delmonico steaks, 251
denuding, 58
    clod heart, 222–23
    outside round, 291–92
    skirt steak, outside, 242
    strip loin, 266–67
    tenderloin, 272–74
Denver steaks, 231
deoxygenated myoglobin, 14
disposal of inedibles, 89, 133
dog bones, 63, 306
dog food
    edible offal and, 85
    natural, 308
dried tissue, 56
drip loss, 313
drugstore method, wrapping meat, 320
dry-aging, 18–19

E

E. coli (*Escherichia coli*), 24, 25
Edge Pro, 40, 41
edge sealers, 316–17, 318
either/or decisions, butchering, 194
elastin, 7
English-style ribs, 239
entrails. *See* evisceration; inedibles; offal
enzymes, 65
equipment. *See also* knife accessories; tools
    bandsaw, 46
    basics, 94, 136
    bench scraper, 31, 39, 42
    bone dust scraper, 42
    boning hook, 35, 42
    butcher's twine, 39, 42–43
    cutting boards, 43–44

grinders, 46
kitchen scale, 44
meat lugs, 47
optional, 46–47, 136
required, 136
spray bottle, 44
tabletop, 42–44
towels, 44
esophagus, 97
removal without weasand rod, 121
separating with weasand rod, 106–7
evisceration, 84. *See also* viscera
abdominal, 117–122
carcass trimming and, 130
hoisting and, 81
thoracic viscera, 123
without a weasand rod, 121
exsanguination, 80–81
blood vessels and, 97
collecting blood/disposing of, 80
hanging during, 83
insensibility and, 77
testing for life/death, 81
thoracic method, 80–81
transverse method, 80, 100
eye of ribeye, 257
eye of round, 208, 286

F

fajita meat, 146, 241, 242
fascia, 57, 60
fascial layer, removal of, 62
fast-twitch fibers, 12, 14, 15
fat(s). *See also* marbling
caul fat, 119
as connective tissue, 9
deposits, types of, 10
intermuscular, 10
intramuscular, 10
kidney fat, removal of, 185
meat tenderness and, 12
saturated, 10–12
shaping and, 57
suet and other, 309
taste and, 9–10
unsaturated, 10–12
visceral, 10
fat caps, 57, 202
feather bones, 245, 255, 265
feed, withholding, 76, 95
femur, 51, 60, 63, 96, 137, 138
femur tendon, 205
fermentation, 65
filet mignon
portioning steaks, 276
tenderloin and, 145
finger bones, 196, 265
first 24 hours postmortem, 16–17
fisting, 83, 84, 114
flanken-style ribs, 240

flank primal, 137, 146, 281–84
cuts, separating main, 188–192
flank steak, 146, 188, 282, 284
removal of, 186–87
sirloin flap, 283–84
trimming, 192
"flap meat," 281
flat iron steak
cutting, 230
described, 227
making, 228–29
flavor of meat
fat cells and, 9–10
off-flavors and, 316, 322–23
slow growth and, 1
wrapping/freezing meat and, 311
flex knife blades, 36, 37
flexor muscle, removal of, 288
"flight zone," 78
foodborne illnesses. *See also* pathogens
food handling and, 22, 23
vegetables and, 23
foodborne pathogens. *See* pathogens
foodborne viruses, 26
food-grade packaging, 316
food-grade twine, 43
food safety. *See also* contamination; pathogens
antibiotics and, 21
microbes and, 22
forequarter, 139–143, 151–57
breaking down, 151–57
brisket/foreshank primal, 141
carcass breakdown, 139
chuck primal, 140
inside/outside skirt, removal of, 152–53
plate primal, 143
quartering and, 132
rib primal, 142
splitting chuck from rib, 154–55
splitting rib and plate, 156–57
veal, breaking down, 151
foreshank. *See* brisket/foreshank primal
foundations, butchering methods
bone saw, finesse with, 51
boning knife, holding, 50
honing rod, sharpening with, 52–53
freezer burn, 136, 313
freezer-grade wrappings, 317
freezers, 32, 312
freezing, 312–15
dehydration and, 313–14
effects of, 312–13
ice crystals and, 313
leaner cuts and, 315
meat quality and, 311
offal, 133, 314
palatability and, 93
pathogens and, 313
portioning and, 314–15
rapid, importance of, 313–14

sanitation and, 314
temperature and, 314
frenching, 42, 62

G

gallbladder, 96
gambrel cords
hanging implement and, 108
hoisting and, 81–82
hooves, removal of, 102
removal of, 300
gambrels, 82
gastrocnemius, removal of, 288
gelatin, 8, 63, 299, 306
genitals, separating, 104
glands
as inedible trim, 56
mammary gland, 104
gloves
cut-resistant, 44
nitrile, 44, 45
grab hooks, 82–83
grain direction of meat
denuding and, 58
tenderization and, 64
trimming and, 54, 56
Grandin, Temple, 77
grinders, types of, 46, 47
grinding meat, edible trim and, 57
grip, boning knife and, 50
ground meat. *See also* hamburger
bags for, 323
contamination and, 22
fat content and, 57
foodborne illnesses and, 33
sirloin tip and, 296

H

hamburger, 33, 64. *See also* ground meat
handsaw. *See* bone saw/handsaw
hand washing
sinks and, 30
thorough, 29–30
hanger steak, 129, 260–61
hanging
carcass breakdown, 139
implement, inserting, 108
quartering carcass and, 97
head
removal of, 124–25
skinning, 125
heart, 86, 96
cardiac muscle, 11
cleaning, 308
removal of, 121, 123
heel, 287–89
seaming, 288–89
shank and, 211–12
hide(s). *See also* skinning
selling, 133

tanning, 125
hindquarter, 144–47, 184–212
    flank, removal of, 186–87
    flank cuts, separating main, 188–192
    kidney fat, removal of, 184
    loin primal, 145, 194–203
    quartering and, 132
    splitting loin from round, 193
    veal, 212
hoisting, 81–83
    hanging implement and, 108
    lifting options, 82
    securing the animal, 82–83
honing rod/knife steel, 91
    finding your angle, 52
    freehand use of, 53
    tabletop method, 53
hooves, removal of, 102
hot boning, 88
housing, humane, 76
humerus, 138
    boning and, 61, 299
    removal of, 164
hydrolysis of collagen, 8
hygiene. *See also* cleanup
    clothing and, 30
    personal, 29–30
    workspace, 22

I

ice glazing, 322
inedibles
    disposal of, 89
    separating, 56–57
infections
    incubation period, 23
    invasive/noninvasive, 22–23
infectious bacteria, 25
insensibility, 77
inside round. *See* top round
inside skirt, removal of, 152–53, 190
inside skirt steak, 143, 146, 241
intercostal meat, 57, 181
intermediate fiber, 15
intermuscular fat, 10
intestines, sausage and, 85
intramuscular fat, 10
invasive infections, 22
involuntary muscles, 11, 12

J

Jaccard meat tenderizer, 46, 47, 65, 223
joint(s)
    atlas, 97, 124, 125, 138
    boning, basics, 60, 62
    break, 138
    carpal, 96, 97
    separating, 62

stifle (knee), 210
tarsal, 96, 97, 102

K

kabob meat, 296
kidney fat, removal of, 185
kidney(s), 86, 122
kitchen scale, 44
knife accessories
    honing rod/knife steel, 38, 39, 40, 52–53, 91
    scabbard, 40, 41
knife steel. *See* honing rod/knife steel
knives. *See also* boning knife
    blade characteristics, 36
    boning knives, 36–38, 39
    breaking knife, 38, 39
    butcher knife, 38, 39
    cimeter, 38, 39
    flexibility of, 36
    shape of, 36
    sharpening, 40, 41
    types of, 34–35, 36–38, 90, 91
knots, making, 70–73
    butcher's knot, 70–71
    packer's knot, 72–73
knuckles, 306

L

labeling, 316, 323
leaner cuts, 315
lifter meat, 244
lifting options, 82
ligament, nuchal, 178
liver, 86
    cleaning, 308
    removal of, 120
liver flukes, 120
locally raised animals, 1
loin, 258–280. *See also* short loin; sirloin; tenderloin
    ball tip, 259
    bottom sirloin butt, seaming, 201
    breaking down, 194–203
        boning the sirloin, 198–200
        short loin, 197
        tenderloin removal (optional), 195–96
    cuts, common, 145
    hanger steak, 260–61
    primal, 137, 145
    protecting, 187
    short, 197
    splitting from round, 193
    steaks, 324–25
        bone-in, 262–263, 268–69
        boneless, 270
    strip loin

        bone-in roast, 264
        boneless roast, 265
        denuding, 266–67
    strip steaks, 268–270
        bone-in, 268–69
        boneless, 270
    subprimals, 145
    top sirloin butt, seaming, 202–3
    top sirloin cap, 277
    top sirloin center, 278–79
    tri-tip, 280
longissimus dorsi
    anterior/posterior faces of, 262
    boning the sirloin and, 200
    eye of ribeye, 257
    most valuable muscle, 145
    rib cuts and, 238
    rib loin and, 156
    in rib primal, 142
    separating, 183
    strip loin and, 264, 267
lungs, removal of, 121, 123

M

machine-assisted cooling, 87
mad cow disease, 28
mallet, 46–47, 65
mammary gland, 104
marbling
    intramuscular fat, 10
    USDA grading system and, 94
marinades, alkaline, 65
marrow
    boning hook and, 42
    roasting, 307
marrow bones, 63, 307
masseter. *See* cheek(s)
meat(s). *See also* flavor of meat; grain direction of meat
    aging, 18–19
    animal stress and, 17
    animal's well-being and, 2, 5, 17
    chemical changes and, 5
    color and freshness, 15, 16
    cured, 15, 315
    damage, avoiding, 83–84
    fajita, 146, 241, 242
    "flap meat," 281
    ground, 22, 33, 57, 64, 323
    intercostal, 57, 181
    leaner cuts, 315
    lifter, 244
    preservation of, 16
    stew, 162, 203, 225, 290
    thawing, 322, 323
    water within, 312
meat grinders, 46, 47, 64–65
meat hammers, 65
meat hooks, 132, 139

meat lugs, 47
meat tenderizer, 46, 47, 65, 223
meat tenderness
    age of animal and, 9
    aging and, 18
    connective tissue and, 12, 18
    enzymatic activity and, 18, 19
    fat and, 12
mechanical tenderization, 46, 47, 64–65
membrane(s)
    skinning and, 84
    trimming and peeling, 152, 189, 209
merlot cut, 287, 289
microbes, 22, 32
"middle meat" primals
    loin primal and, 145
    rib primal and, 142
mobile slaughter truck, 81
mouse muscle, 203, 279
mouthfeel, 64
multifidus dorsi, separating, 183
muscle(s), 10. *See also* grain direction of meat; heart; *longissimus dorsi*
    about, 6
    accessory, 203
    age of animal and, 7
    chemical changes and, 5
    color of, 13–16
    connective tissue, 7–8
    denuding, 58
    fats and, 10
    fibers, 6–7, 12–13
    function of, 6–7
    into meat, 16–17
    mouse muscle, 203, 279
    pH of, 17
    seam, 49, 61, 62
    seam cutting, 59
    structure, 6, 7, 49
    tenderness, collagen and, 8, 9
    types of, 11, 12
myoglobin
    cured meats and, 15
    oxygen exchanges and, 13
    stages of, 14–15

N

nature-assisted cooling, 87
navel, 236–37
    boned and rolled, 237
    "danger zone" and, 24
neck, 224
    boning, 174–75
    removal of, 60
neck bones, 173, 306
nitrile gloves, 44, 45
noninvasive infections, 22–23
norovirus, 26
nuchal ligament, 178

O

offal. *See also* inedibles
    candidates for, 97
    edible, 85–86
    freezing, 133, 314
"on the hook." *See* hanging
organs. *See* offal
outside round, 209, 290–93
    denuding, 291–92
    western tip/western griller steaks, 293
outside skirt, removal of, 152–53
outside skirt steak, 143, 242
oxidized myoglobin, 14
oxtail, 85, 309. *See also* tail
oxygenated myoglobin, 14
oyster steak, removal of, 205

P

packaging. *See also* wrapping meat
    about, 136
    airtightness, 315
    durable, food-grade materials, 316
    expulsion of interior air, 315
    ice glazing, 322
    labeling and, 316
    materials, additional, 322–23
    meat quality and, 311
    solutions, 316–322
    vacuum sealing, 136, 316–17, 318
    wrappings, freezer-grade, 317
    zip-top bags, 322
packer's knot, 72–73
pale, soft, and exudate (PSE) meat, 17
pancreas, 56
paracord, 82
parasites
    liver flukes, 120
    parasitic worms, 26–27, 32, 33
patella (kneecap), 210
pathogens. *See also* foodborne illnesses
    antibiotics and, 21
    bacteria, 24–25
    cold storage and, 32
    contamination and, 22–23
    foodborne, 22–23
    freezing and, 313
    "industrial strength," 21
    infections and, 23
    killing with heat, 33
    preventing spread of, 29–31
    tools and, 30
    types of, 24–28
pectinius, removal of, 302
peeling the bone
    fascial layer, removal of, 62
    step-by-step, 63
pelvic girdle, 137
pelvis. *See* aitchbone

personal hygiene. *See* hygiene
pH of muscles, 17
pistol grip, 50
pizzle, 104, 109
plate primal, 235–242
    described, 137
    navel, 236–37
    short ribs, 238, 239–240
    skirt steak, inside, 241
    skirt steak, outside, 242
    splitting rib and, 156–57
pluck, removal of, 121, 123
poisonous bacteria, 25
porterhouse steaks, 263, 325
    cut sheet and, 69
    either/or decisions and, 194
    tenderloin and, 150
portioning
    based on thickness, 67
    based on weight, 66
    basics, 65
    prior to freezing, 314–15
    shaping and, 57
postmortem, first 24 hours, 16–17
pre-slaughter conditions
    day-before measures, 76
    humane conditions, 75, 76
    separating animals, 76, 95
    withholding feed, 76
pre-slaughter setup
    age of animal and, 94–95
    preparation for slaughter, 95
pre-slaughter stress, 17
primal(s). *See also* subprimals
    described, 137
    most valuable, 194
prions, 28
projectile stunning, 78, 79
protective aprons, 45
proteolysis, 18
protists, 27–28
    life cycle of, 28
    Toxoplasma gondii, 28
    transmission, prevention of, 28
PSE meat, 17
pubic symphysis
    splitting, 113
    tying off the bung, 113

Q

quadrupeds, 137, 271
quartering, fore- and hindquarters, 132

R

ranch steaks, 223
refrigeration, 32
rendering, 57, 89
rhomboideus, 179, 231
rib(s)
    back, 238, 255
    cuts, common, 142, 238
    English-style, 239
    flanken-style, 240
    and plate, splitting, 156–57
    and plate primals, 137
    short, 143, 176, 239–240
        chuck, 225
    splitting, 131
    splitting chuck from, 154–55
ribeye
    cap steak, 252, 256
    roll, 253–57
        cutting, 254–55
        rib cover removal, 253
        seaming, 256–57
rib loin, 156
rib primal, 142, 243–257. *See also* ribeye
    rib roast
        bone-in, 244–47
        bone-on, 248–49
    rib steaks
        bone-in, 250
        boneless, 251
rigor mortis, 16–17, 18
roast(s)
    ball tip, 259
    under blade roast, 181, 231
    butcher's knot and, 70–71
    packer's knot and, 72–73
    rib
        bone-in, 244–47
        bone-on, 248–49
    strip loin, 264–67
        bone-in, 264
        boneless, 265
        denuding, 266–67
    from top round, 305
"rose" veal, 95
rotovirus, 26
round. *See also* bottom round; top round
    bottom, separating muscles of, 208–9
    breaking down, 204–12
    cuts, common, 147
    eye of, 208, 286
    heel, 287–89
        and shank, 211–12
    outside, 290–93
    oyster and aitchbone, removal of, 205
    primal, 137, 147, 285–305
    sirloin tip, 294–97
        peeling, 210
    splitting loin from, 193
    subprimals, 147
rumen, 96
    carcass trimming and, 130
    caul fat and, 119
    cleanup and, 133
    esophagus and, 106, 121
    manure, 89
    splitting brisket and, 105
    tripe, 85, 97

S

sacrum, separating, 199
salmonella, 25, 27
sanitary clothing, 30
sanitizing
    freezing and, 314
    quick dip, 30–31
    solution, 31
sausage, 46, 47, 54, 64
    cut sheet and, 68
    edible containers for, 85
sausage stuffer, 47
saw. *See* bone saw/handsaw
scabbard, 40, 41
scale
    kitchen, 44
    packaging and, 323
scapula
    cut points and, 138
    removal of, 172, 245
scimitar, 38
scrap meat, 306
sealers
    chamber, 317, 319
    edge, 316–17, 318
seam, muscle, 49, 61, 62
seam cutting, 59
semi-stiff knife blades, 36, 37
sensibility, 77, 83
setup for butchering, 136
shank, 298–300
    boning, 299
    cooking, 8
    cross-cut, 300
    heel and, 211–12
    hydrolysis of collagen and, 8
shank meat, 141
shaping, edible trim from, 57
sharpening knives. *See* honing rod/knife steel
short loin, 197
    cut sheet and, 69
    as loin subprimal, 145
    splitting, 131
short plate, 238
short ribs, 143, 176, 238
shoulder clod, 140
    and chuck tender, separation of, 165–69
    seaming out, 170–71
    steak, 221
shoulder tender, 170, 226
"siding"/side hide removal, 103
silverside. *See* outside round
silverskin, removal of, 58, 170, 247, 292
singletree, 82
sinks, handwashing and, 30
sirloin, 145. *See also* top sirloin
    boning, 198–200
    butt, seaming
        bottom, 201
        top, 202–3
    cap, 202
    flap
        cleaning, 283–84
        finding, 186
        peeling top of, 190
    tip, 147
        peeling, 210
        seaming, 295–97
skeletal muscle, 11
skeletal structure. *See* anatomy/skeletal structure
skinning, 83–84. *See also* hide(s)
    beginning to remove hide, 103
    "clean hand"/"dirty hand," 83, 84
    contamination and, 23, 83–84
    finishing, 116
    fisting, 83, 84, 114
    head, 125
    hoisting and, 81
    meat damage and, 83–84
    round and rump, 109
    situating carcass for, 101
    wet carcass, dry hide, 84
slaughter
    animal stress and, 17
    postmortem, first 24 hours, 16–17
    process, flowchart, 98
    rigor mortis and, 16–17, 18
    storage after, 17
slaughter truck, mobile, 81
slow-twitch fibers, 12–13, 14, 15
smooth muscle, 11, 12
space requirements, 91
spinalis dorsi, separating, 183
spleen, 85, 86
spoilage bacteria, 24–25, 32
spray bottle, 44
square-cut chuck, 158–172
    boning, 61
        neck, 174–75
    breakdown
        quick, 162
        seamed, 163
    brisket and foreshank
        removal of, 159–160
        separation of, 161
    chuck eye roll, seaming, 182–83
    chuck roll
        boning, 176–78
        boning and seaming, 173–183
        seaming, 179–181

humerus, removal of, 164
primal, 137
scapula, removal of, 172
shoulder clod
    and chuck tender, separation of, 165–69
    seaming out, 170–71
standing rib roast, 244–47
steak(s)
    ball tip, 259
    blade, 178, 217
    cutting, grain direction and, 64
    Denver, 231
    filet mignon, 145, 276
    flank, 146, 188, 282, 284
    flat iron, 227–230
    hanger, 129, 260–61
    oyster, 205
    porterhouse, 69, 150, 194, 263, 325
    ranch, 223
    rib
        bone-in, 250
        boneless, 251
    ribeye cap, 252, 256
    shoulder clod, 221
    skirt
        inside, 143, 146, 241
        outside, 242
    strip
        bone-in, 268–69
        boneless, 270
    T-bone, 69, 194, 263, 324
    top blade, 227
    from top round, 305
    "vein steaks," 263, 267
    western tip/western griller, 293
steer. *See* animal
sternum, 96, 97
    boning, 233
    sawing through, 105, 155
stew meat, 162, 203, 225, 290, 296
sticking knife, 38
stiff knife blades, 36, 37
stifle joint (knee)
    cutting through, 212
    patella and, 210
stock bones, 306
storage of inedibles, 89
storage of meat. *See also* freezing
    after slaughter, 17
    leaner cuts, 315
striated muscle, 11, 12
strip loin
    bone-in roast, 264
    boneless denuded, 55
    boneless roast, 265
    denuding, 266–67
    portioning
        based on thickness, 67
        based on weight, 66
strip steaks, cut sheet and, 69

stunning, 76–79
    basics, 99
    captive bolt, 77, 78, 79, 90
    insensibility and, 77
    projectile, 78, 79
    spot, locating, 79
subcutaneous fat, 10
subprimals. *See also* primals
    cuts and, 214–15
    described, 136
suet, 309
surfaces. *See* cutting boards
sweetbreads, 56, 86
    veal, 123

T

tabletop equipment, 42–44
tail. *See also* oxtail
    bones in, 110
    length of, 187
    removal of, 111–12
tallow production, 309
tapeworms, 27
tarsal joint, 96, 97, 102
T-bone steaks, 263, 324
    cut sheet and, 69
    either/or decisions and, 194
temperature
    carcass cooling, 133
    cold storage and, 32
    "danger zone," 23–24, 32
    freezing and, 313, 314
    hydrolysis of collagen and, 8
    thawing meat, 323
tenderization
    chemical, 65
    cutting across the grain and, 64
    mechanical, 46, 47
tenderloin, 271–76
    cut sheet and, 69
    denuding, 272–74
    either/or decisions and, 194
    filet mignon and, 145
    porterhouse steaks and, 150, 194
    removal of (optional), 195–96, 258
    silverskin and, 58
    steaks, portioning, 276
    trimming and, 56
    tying, 275–76
tenderness. *See* meat tenderness; tenderization
tendons. *See also* gambrel cords
    femur, 205
    as unpalatable trim, 57
testicles, removal of, 104
thawing meat, 32, 322, 323
thaw shortening, 17
thoracic method, exsanguination, 80–81
thoracic viscera, removal of, 123
thymus gland, 56, 123
tongue, 86, 309

tools. *See also* equipment; knives
    about, 90–91
    basics, 34–35
    bone saw/handsaw, 38, 41, 51
    cleaver, 39, 40–41
    mallet, 39, 46–47
    sanitizing, 30–31
    work area and, 89
top blade, 166–67, 171, 227
top round, 301–5
    cap, 304
    portioning, 305
    round primal and, 147
    seaming, 302–4
    separating, 204–5
top sirloin
    butt, seaming, 202–3
    cap, 277
    center, 278–79
towels, 44
Toxoplasma gondii, 28
trachea
    esophagus and, 97, 106
    removal of, 123
transverse method, exsanguination, 80, 100
trapezius, removal of, 179
trash, 89
Trichinella spiralis, 26–27
trim
    edible, from shaping, 57
    inedible, removing, 56
    unpalatable, removing, 57
trim hooks, 91
trimming, 54–59
    carcass, 130–32
    with the grain, 54, 56
    as little as possible, 54
    start to finish, 55
    tenets of, 54, 56
tripe, 85, 97
tri-tip
    cleaning/portioning, 280
    separating, 201
tropocollagen fibers, 9
Tucson cut, 301, 305
twine. *See* butcher's twine; paracord

U

unctuous gelatin, 63, 299
under blade roast, 181
under blade flap, 178
unpalatable trim, 57

V

vacuum sealing, 136, 316–17, 318
variety meats, 308–9
veal
    breaking down, 148–49
    breast, 148
    carcass, splitting, 149

caul fat, collection of, 119
chops, 51, 148, 212
cooling and aging, 133
fell, amount of, 114
fisting, 83, 114, 115
forequarter, breaking down, 151
hindquarter, breaking down, 212
incision map, 115
keeping whole, 128
"rose," 95
sweetbreads, 123
veal calves
    average weight and, 96
    humane conditions and, 95
vegetables
    cold storage and, 32
    foodborne illnesses and, 23
"vein steaks," 263, 267
vertebrae
    removal of, 177
    as stock bones, 306
viruses, 26

viscera, 84. See also abdominal viscera, removal of; evisceration
    cleanup and, 133
    skeletal structure and, 96–97
    thoracic, removal of, 123
visceral fat, 10
voluntary muscles, 11, 12

W

walk-in coolers, 87
weasand rod
    separating esophagus with, 106–7
    viscera removal without, 121
weight. See also scales, cow/calf, average, 96
western tip/western griller steaks, 293
wet-aging, 19
whole cuts, cooking, 33
withholding feed, 76, 95
woods, butcher blocks, 43–44
work area and tools, 89
working-muscle groups, 140

workspace hygiene, 30–31
work surfaces, carcass size and, 43
worms, parasitic, 26–27, 32, 33
wrapping meat. See also packaging
    butcher method, 321
    drugstore method, 320
    foil as layer in, 322
    freezer-grade wrappings, 317
    materials, additional, 322–23
wrappings, freezer-grade, 317

Y

yield
    average, 63
    bones and, 63
    cut sheet and, 58
    maximum, 2, 57, 68

Z

zero-waste approach, 57

Build Your Barnyard Skill Set with More Books from Storey

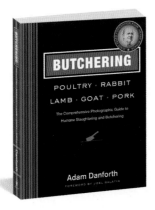

BY ADAM DANFORTH
Learn how to slaughter and butcher small livestock humanely with the help of hundreds of detailed step-by-step photos. This award-winning guide covers food safety, freezing and packaging, tools and equipment, butchering methods, and preslaughter conditions.

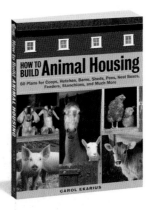

BY CAROL EKARIUS
This all-inclusive guide contains illustrated diagrams and detailed explanations for building 60 shelters that meet animals' individual needs, including barns, windbreaks, and shade structures, plus watering systems, feeders, chutes, and stanchions.

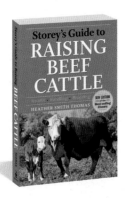

BY HEATHER SMITH THOMAS
This bestseller is the classic cattle-raising reference. With detailed information and expert advice on breeds, behavior, fencing, housing, health care, and more, you'll find everything you need to help your herd thrive.

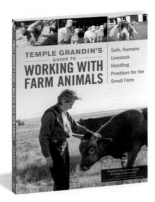

Keep your animals calm and safe with Dr. Grandin's groundbreaking methods, tailored for small farms. This in-depth manual describes the behavior, fears, and instincts of herd animals and teaches you how to set up the most humane and productive facilities on your farm.

Join the conversation. Share your experience with this book, learn more about Storey Publishing's authors, and read original essays and book excerpts at storey.com.
Look for our books wherever quality books are sold or call 800-441-5700.